Applied Electronics

Applied Electronics

JOHN C MORRIS

BA, IEng, FIEIE, CertEd

Senior Lecturer
Department of Engineering
Havering College of Further and Higher Education

ARNOLD

A member of the Hodder Headline Group
LONDON • SYDNEY • AUCKLAND

First published in Great Britain in 1997 by
Arnold, a member of the Hodder Headline Group,
338 Euston Road, London NW1 3BH

Whilst the advice and information in this book is believed to be true and
accurate at the date of going to press, neither the author nor the publisher
can accept any legal responsibility or liability for any errors or omissions
that may be made.

British Library Cataloguing in Publication Data
A catalogue record for this book is available from the British Library

ISBN 0 340 65284 5

Typeset in 10.5pt Garamond by
Paston Press Ltd, Loddon, Norfolk
Printed and bound in Great Britain by The Bath Press, Bath

For Ian and Adam,

'*All experience is an arch, to build upon.*'
Henry Adams

CONTENTS

—

PREFACE

—

The world of electronics is fast changing and may appear a confusing and abstract collection of devices and circuits. There are, however, few areas and aspects of everyday life that are not affected in some way by 'electronics'. Cars contain more electronic circuitry than ever before and many 'electrical' appliances now have additional facilities and improvements that are, by nature, electronic.

To meet the increasing challenge of electronics, today's engineer and technician must be able to demonstrate both a theoretical and practical mastery of the subject.

In order to achieve this 'marriage' between knowing and doing, a 'student centred' step-by-step approach has been adopted that will encourage a familiarity with devices, circuits and applications. This book covers the analogue and digital aspects of electronics by introducing the semiconductor devices themselves, progressing through amplifiers and power supplies to finish with combinational and sequential logic circuits.

Theory is supported by manufacturers' data sheets, worked examples and 37 practical investigations. Only a basic underpinning knowledge is required to develop a thorough understanding of electronic circuits, concepts and design procedures. To help the reader monitor progress, self-assessment questions are included throughout with additional multiple choice questions provided at the end of each chapter.

The material is ideally suited to the 'new' Advanced GNVQ Engineering courses and meets all the element requirements of unit 13 'Advanced Electronics'. The principal objectives of BTEC Electronics courses at level NII are also covered and there is much common ground with City and Guilds and GCSE courses.

Each chapter includes a review section to enable the reader to recap the important points without having to hunt through the text. It is my intention that this book will serve as a 'user friendly' guide to the subject and consequently should be of interest to enthusiasts, technicians and teachers as well as the student of electronics.

The pages within represent the collective efforts of a number of people. I would like therefore to thank Farnell Components Ltd and RS Components Ltd for permission to reproduce extracts from their catalogues and data sheets. My colleagues, as always, have given enormous support with their kindness, wit and many helpful suggestions.

Finally, I should like to extend grateful thanks to my family; to Ian and Adam for their patience and understanding and to Lin, my wife, for her constant help and warm encouragement throughout this project.

John C Morris
Billericay, Essex. September 1996

INTRODUCTION

—

The chapters are presented in a sequence that is suitable for use as a learning programme; however, each topic is separate and can readily be studied in isolation if desired.

All the circuits are proven and are designed so that preferred values of resistors and capacitors can be used. The suggested semiconductors are merely suggestions and comparable or equivalent components can be selected from the data sheets provided, or from a suitable catalogue.

The equipment required is of the type to be found in any basic electronic laboratory or repair workshop. For safety reasons all a.c. circuits are designed to operate at a low voltage, supplied via a mains step-down transformer. Under no circumstances should these specified voltages be exceeded. The method of circuit construction is left to the reader but the use of a breadboard is highly recommended since it offers speed of construction and allows components to be reused.

Students following GNVQ courses have to provide evidence of practical skills. The investigations provided in this book encourage the development of such skills. The majority of the practical investigations are easy to do and quick to complete so it is envisaged that most of the learning will be via a practical 'hands on' approach. The more detailed investigations can be modified and expanded to develop important report writing and presentation skills that have such relevance in today's engineering world.

When carrying out the investigations, work methodically and pay attention to the following points.

1 Check that the circuit is correct *before* switching on the supply.
2 Check the circuit voltage with a meter when setting to a specified value (meters on power supplies are there for guidance only!).
3 Make sure you record *everything*.
4 Plot graphs *as you go*.
5 *Do not* dismantle your circuit immediately you have finished but check through what you have written and what is yet required so that if you have to repeat part of the procedure the anger and frustration will be minimised.

SEMICONDUCTOR DIODES

—

What exactly is a semiconductor?
To answer this question let us first consider a conductor and an insulator. To do this, a quick look at basic atomic theory is required.

The nature of matter

All matter can be broken down until the smallest part is arrived at – the *atom*. An atom consists of a nucleus at the centre around which electrons revolve or orbit (Fig. 1.1).

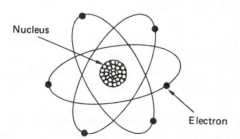

Nucleus

Electron

Fig. 1.1 The atom

An electron is bound to its atom because the nucleus has a positive charge and the electron a negative charge $(1.6 \times 10^{-19}$ Coulombs), so a force is exerted on the electron and it follows a spherical orbit about the nucleus.

The number of electrons an atom possesses depends upon the material. Hydrogen has one orbiting electron and helium has two. These are simple elements. However, more complicated materials have many electrons. These electrons

may not share the same orbit but revolve in layers called *shells*, as shown in Fig. 1.2.

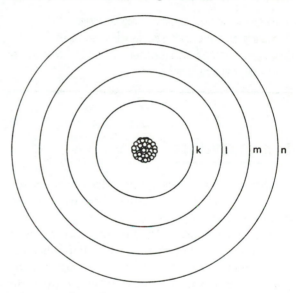

k l m n

Fig. 1.2 Electron shells

The electrons in the inner shells are very tightly bound to the nucleus but those on the outer shell are not so tightly bound. The electrons on the outermost shell are called *valence electrons*.

Current flow

Current flow is the movement of charge carriers in a material. In other words, for metals like copper it is the movement of electrons. *How?* Valence electrons are quite easily dislodged from their orbit and then are free to move within the material. To do this, a force must be exerted on the electron in the form of a voltage or *potential difference*.

WHAT IS A CONDUCTOR?

A conductor is a material that has a large number of 'free' electrons, like most metals, e.g. copper, silver, aluminium. Consequently, a conductor has a very low resistance and current will flow easily.

WHAT IS AN INSULATOR?

A material with few free electrons, e.g. rubber, PVC, paper. All insulators have a very high resistance and current will not flow easily.

SELF ASSESSMENT 1

1 Does an electron have a positive or negative charge?
2 Where in the atomic structure are the valence electrons?
3 If a piece of material had a resistance of $3 \times 10^{-8}\ \Omega$ would it be an insulator or a conductor?

Semiconductors

As their name suggests, semiconductors are materials that have characteristics that are between conductors and insulators. There are a number of semiconductor materials, but the most common are silicon (Si) and germanium (Ge). Pure silicon or germanium is called an *intrinsic* semiconductor and possesses four valence electrons (electrons in the outermost shell). It is said to be a *tetravalent* material (*Tetra* meaning four).

Intrinsic silicon or germanium has a crystalline lattice structure with all the atoms arranged in an orderly and regular pattern. The four valence electrons share their orbit with four neighbouring atoms.

In this way the atoms are bonded together to form the lattice structure. This sharing of orbits is called *covalent bonding* and although there is continual movement of electrons between atoms, at no time does any atom have more or less than four valence electrons (Fig. 1.3).

CONDUCTION IN INTRINSIC SEMICONDUCTORS

The bonding process appears to use all valence electrons leaving no free ones for conduction. This is true at very low temperatures, but at room temperature some valence electrons acquire sufficient thermal energy to break free of their bonds and are therefore available for conduction.

What happens when a bond breaks? The electron leaves behind it a *hole* and, since the electron has a negative charge, the hole has a positive charge: an *electron–hole pair* has been generated.

In a pure semiconductor there are exactly the same number of holes as free electrons and the electrons can move from hole to hole. When an electron enters a hole, recombination has occurred.

- Would current flow if a potential difference was applied to a piece of pure silicon?
- Are there any free electrons? Depends on the temperature.
- OK. It is at 20°C (room temperature). Then there are free electrons.
- Are there any holes? Yes the same number as free electrons.
- Will these move? Let's see in Fig. 1.4.

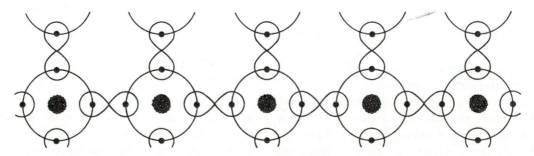

Fig. 1.3 Covalent bonding

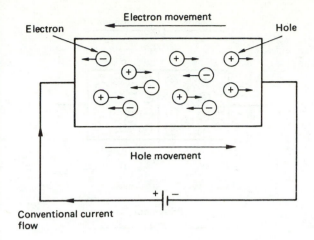

Conventional current flow

Fig. 1.4 Movement of charge carriers

How can a hole move?

It is actually the electrons that move – consider a dentist's waiting room (Fig. 1.5). You can see that it is an effective movement of holes caused by the movement of electrons.

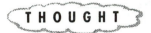

THOUGHT

■ *If current flow is electron flow what then is conventional current flow? Before the electron was discovered, scientists needed to explain the phenomenon of current flow. It seemed reasonable (at the time) that current should flow from positive to negative and this convention was adopted. After the discovery of the electron it become obvious that electron flow was from negative to positive so today we have to accept the two.*

1 Electron flow is from negative to positive.
2 Conventional current flows from positive to negative.

In practice this rarely causes problems because we tend to think in terms of one or the other but not both.

EXTRINSIC SEMICONDUCTORS

Pure silicon (or germanium) can have its conduction characteristics changed by the introduction of an impurity. It then becomes known as an *extrinsic* semiconductor of which there are two types; n-type and p-type.

n-type semiconductor

Pure silicon (or germanium) is tetravalent (four valence electrons). If an impurity material is introduced, the atoms of which have five valence electrons (a *pentavalent* material) then atom for atom they will fit into the lattice structure, but now for every impurity atom there will be five electrons where only four are required. So an excess of electrons will exist and the material becomes n-type silicon or germanium. This introduction of an impurity is called *doping*. The common pentavalent doping materials are arsenic, phosphorous and antimony.

p-type semiconductor

If the pure semiconductor is doped with an impurity material that has only three valence electrons (*trivalent* material) then for every impurity atom there will only be three valence electrons where four are required: this results in a deficiency of electrons or a surplus of holes. Common trivalent doping materials are aluminium, gallium and boron.

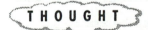

THOUGHT

■ *Has n-type silicon got a negative charge? No. An n-type semiconductor is not negatively charged, it simply has a surplus of electrons which are negative charge carriers. Likewise p-type material has a surplus of holes. Therefore it has positive charge carriers.*

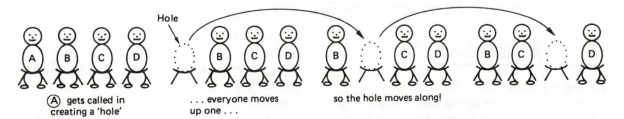

Fig. 1.5 Electron/hole movement

MAJORITY AND MINORITY CHARGE CARRIERS

In n-type semiconductors electrons are extremely numerous and are the *majority charge carriers*, but there will always be a few holes present due to thermal generation or unwanted impurities. These are the *minority charge carriers*. In p-type semiconductors the minority charge carriers are electrons with holes forming the majority charge carriers.

CONDUCTION IN EXTRINSIC SEMICONDUCTOR

When a piece of n-type or p-type silicon (or germanium) is connected to a potential difference the charge carriers will move. Fig. 1.6 shows the movement of charge carriers in n-type material.

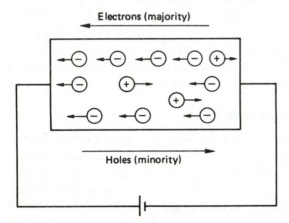

Fig. 1.6 Movement of charge carriers in n-type silicon

SELF ASSESSMENT 2

1 How many valence electrons in tetravalent germanium?
2 Will pure silicon conduct at room temperature?
3 If a piece of intrinsic germanium at 50°C was said to have 3×10^{28} 'free' electrons, how many 'holes' would have been generated?
4 Sketch a piece of p-type germanium connected to a potential difference, identify the majority and minority charge carriers and show their relative movement.

The p–n junction diode

The diode is a one-way street for current flow
If a piece of silicon or germanium is doped so that half is n-type and half p-type then a p–n junction diode is formed (Fig. 1.7(a)).

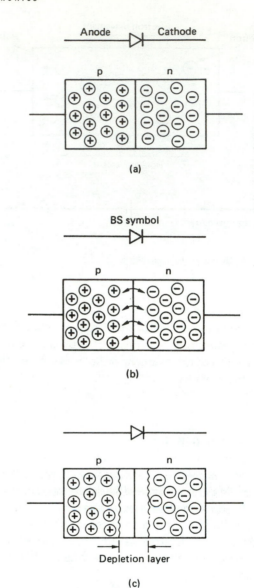

Fig. 1.7 The p–n junction diode

At the instant of manufacture, charge carriers, in the region of the junction move across and recombine (Fig. 1.7(b)). This results in regions close to the junction that are depleted of majority charge carriers. These regions are called the *depletion regions* or *layers* (Fig. 1.7(c)). So, in the p-type material near the junction is a region that is less positive, i.e. negative, and in the n-type material there is an area that is less negative, i.e. positive. Therefore a potential barrier exists, the size of which is determined by the type of semiconductor and the type of diode construction, but is approximately 0.6V for silicon and 0.3V for germanium. Let us now examine what happens when the diode is connected to a potential difference.

THE REVERSE BIAS DIODE

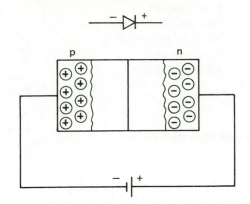

Fig. 1.8 The reverse bias diode

When connected as shown, the negative charge carriers (electrons) are attracted to the positive potential and the positive charge carriers (holes) to the negative potential. The depletion layer widens, the resistance of the device is very high and no current flows.

THE FORWARD BIAS DIODE

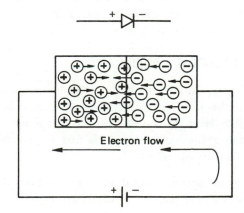

Electron flow

Fig. 1.9 The forward bias diode

When connected as shown, the positive carriers move away from the positive potential and the negative carriers are repelled by the negative potential, and so they cross the depletion layer and

recombine. The resistance of the device falls and current can flow.

The practical diode

1 When a diode is forward biased, the depletion layer potential must be overcome before significant current will flow. The anode must be about 0.6V more positive than the cathode for a silicon device and about 0.3V for a germanium device.

2 You can see that the British Standard (BS) symbol for a diode indicates the direction conventional current flows through it (Fig. 1.10).

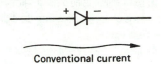

Conventional current

Fig. 1.10 Conventional current direction

3 When reverse biased, the diode has a very high resistance but some current does flow due to the minority carriers. This is called the *leakage current*. This current is so small for a silicon device that it is difficult to measure, but for a germanium device it is higher. Now carry out Practical Investigations 1 and 2.

THE DIODE CHARACTERISTIC

Fig. 1.11 shows that when the anode is positive with respect to the cathode, current will flow once the *barrier potential (B)* has been overcome.

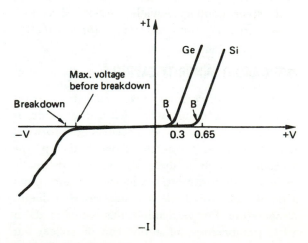

Fig. 1.11 Diode characteristics

PRACTICAL INVESTIGATION 1

The forward bias diode

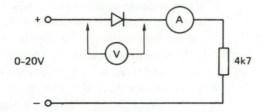

Equipment
Silicon signal diode
Silicon power diode
Variable d.c. power supply
Digital multimeter
4k7 resistor (4.7 kΩ)

Method
1 Set the circuit up as shown using a silicon signal diode.
2 Monitor the voltage across the diode using the digital multimeter and adjust the power supply so that the meter indicates 0.45V.
3 Measure the diode current.
4 Repeat for the range 0.4–0.7V and record your results in a table as shown below.

	V	0.4	0.45	0.5	0.55	0.6	0.65	0.7
Signal diode	I							
Power diode	I							

Results
1 Plot the forward characteristics of both diodes (on the same axis).
2 Calculate the diode resistance at 0.6V for both diodes.

When reverse biased, virtually no current flows (apart from leakage). *But* if the reverse bias voltage is increased, there comes a point where the diode physically breaks down and then current will flow. However, the device will now be destroyed and will either conduct equally in both directions (short-circuit) or not conduct at all (open-circuit).

MORE ABOUT MINORITY CARRIERS

You have discovered that minority carriers increase with temperature when the diode is reverse biased, but this also occurs under forward bias conditions. So as the diode heats up, the current through it increases and the thermally generated minority carriers increase. There comes a point when the diode will stop working due to temperature. For germanium this occurs at about 80°C (temperature of a hot cup of coffee) but silicon devices can continue working up to about

160°C. The importance of keeping a device cool (particularly power diodes) is obvious and that is why *heat sinks* are often used.

HOW TO CHOOSE A DIODE

By glancing at the diode specification sheets you can see that there are many different types of diode available. How do you choose one that is suitable for your application? You must first ask yourself the following questions.

1 Does the application call for a signal diode or a power diode? (i.e. low or high current)
2 What is the maximum forward current that will flow through the diode?
3 What is the maximum forward voltage to be applied to the diode?
4 What is the maximum reverse voltage that the diode must be able to stand before breaking down?

PRACTICAL INVESTIGATION 2

The reverse bias diode

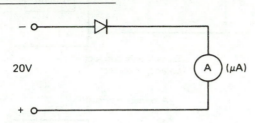

20V

(A) (μA)

Equipment
Silicon signal diode
Germanium diode
Sensitive ammeter (μA)
Thermometer
Hair dryer
d.c. power supply

Method
1 Connect the circuit as shown using a general purpose germanium signal diode (e.g. OA47, OA90, OA91).
2 Set the power supply to 20V and record the reverse bias current (typically 5–10μA).
3 Warm the diode with the hair dryer and observe how the reverse bias current changes.
4 Measure and record the leakage current over the temperature range 20–50°C.

T (°C)	20	25	30	35	40	45	50
I_{rev}							

5 Repeat the above for a silicon signal diode.

Results
1 Compare and comment on the reverse bias current of silicon and germanium diodes.
2 Which diode appears to be more temperature dependent?

Example 1
A diode is required to be used on 40V d.c. carrying 60mA. From the data provided select three suitable diodes.

How do we use the manufacturers' data?
First an understanding of the common terms is required:

I_F – Maximum d.c. forward current.
$I_F(AV)$ – Maximum d.c. current with an alternating component.
V_F – d.c. voltage drop across the diode.
V_{RRM} – The maximum reverse bias voltage that the diode can continually stand before breakdown occurs.
PIV – Peak inverse voltage which is the same as V_{RRM}.

So for Example 1:

I_F = 60mA
V_{RRM} = 40V
V_F – This will be determined by the diode chosen.
$I_F(AV)$ – Not applicable because there is no alternating component present for d.c.

As you will have found, there are many suitable diodes! In fact any diode that has values equal or greater than your requirements will do. Your ultimate choice will probably be determined by preference for silicon or germanium, available stock and delivery time and cost.

Note You may have noticed that although we assume the forward voltage drop to be 0.6 and 0.3V for Si and Ge respectively, in practice there is a considerable variation in V_F according to the type of diode.

Example 2
You are building a half-wave rectifier to operate from the 240V mains and pass a maximum current of 12A. Select a diode.

$I_{F(AV)}$ = 12A
V_{RRM} = CAREFUL! 240V a.c. is the r.m.s. value of the mains voltage. Peak will be $240 \times 1.414 = 340V$

So

$$V_{RRM} = 340V \text{ or greater}$$

A BYX42–600 would be suitable with a 12A rating and 600V V_{RRM}. Note the physical structure: the cathode has a thread or stud, with a nut provided so that it can be bolted to a heat sink in order to dissipate the heat generated when the current flows through it.

Note Make sure you understand which is the anode and cathode of the diode. Generally, for small devices, the cathode is indicated by a band at one end (Fig. 1.12), but for stud-mounted power devices the stud may be either the cathode or anode! (Usually the anode.)

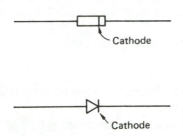

Fig. 1.12 Diode cathode identification

THE CHOICE BETWEEN SILICON AND GERMANIUM

Germanium	*Silicon*
Voltage drop $\simeq$ 0.3V	Voltage drop $\simeq$ 0.6V
Maximum temperature 80°C	Maximum temperature 160°C
Moderate leakage current	Very low leakage current

A quick look at the data sheets will show you that silicon diodes are today more common than germanium ones.

Now use your understanding to study the following circuits and then carry out practical investigations.

Practical diode applications

PROTECTION CIRCUITS

Polarity protection

Many electronic circuits will be damaged should the d.c. be connected the wrong way round (Fig. 1.13).

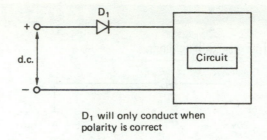

D_1 will only conduct when polarity is correct

Alternative circuit

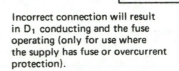

Incorrect connection will result in D_1 conducting and the fuse operating (only for use where the supply has fuse or overcurrent protection).

Fig. 1.13 Polarity protection

Overvoltage protection (*Against back EMF*)

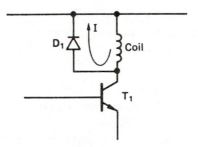

Fig. 1.14 Protection against back EMF

At switch off, a high back EMF could be generated that would damage T_1. D_1 conducts and the back EMF energy is dissipated. This is called a *fly-wheel* diode circuit (Fig. 1.14).

Negative voltage protection

D_1 conducts when the negative spike exceeds 0.6V thus protecting T_1 (Fig. 1.15).

Testing diodes

Because a diode should only conduct one way it is often required to determine whether or not a diode is servicable. This can be achieved using a simple ohmmeter such as a multimeter (e.g. Avometer).

Note The instrument's internal battery is used to bias the diode. However, most instruments have this battery connected so that the positive plate is the negative instrument terminal and the negative plate the positive terminal. Consequently the diode is forward biased when the negative instrument lead is attached to the anode and the positive to the cathode.

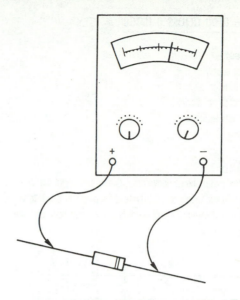

Equipment
Multimeter
An assortment of silicon
and germanium diodes

Method
1 Set the multimeter to the ohms range.
2 Measure and record the forward and reverse resistance of the diodes.

Diode type	Forward	Reverse
IN 914	1.2 kΩ	∞

Note If you measure the resistance of the same diode with different meters the results may not be the same. *Why?* This is because if the internal battery in the instrument has a different voltage the current through the diode will change and hence its resistance will be different. Refer to Practical Investigation 1.

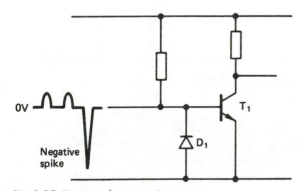

Fig. 1.15 Negative voltage protection

The diode as a rectifier

This is perhaps one of the most common applications of the diode; rectification is the process of converting alternating current to direct current.

THE HALF-WAVE RECTIFIER (Fig. 1.16)

D_1 will conduct only during the positive half-cycles.

Note When considering any circuit with input and output waveforms they should always be drawn one below the other on a common time scale.

THOUGHT

■ *Surely d.c. is a straight line?*
A d.c. voltage is one that has a single polarity. Therefore 0V to positive and back to 0V is d.c.
■ *What is the frequency of the output voltage if the input signal is 50Hz? Answer – 50Hz.*
This circuit then provides unsmoothed d.c. that can be used for certain applications such as battery chargers.

PRACTICAL INVESTIGATION 4

Voltage drop across a diode

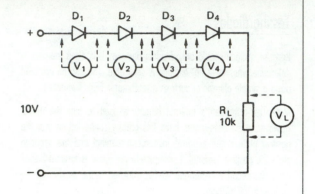

Equipment
d.c. power supply
Digital voltmeter
Four silicon signal diodes
Four germanium diodes
10k resistor

Method
1 Build the circuit shown above using germanium signal diodes.
2 Set the power supply to 10.00V (use the meter to check this).
3 Measure and record the voltage drop across each diode and across the load resistor.

	V_1	V_2	V_3	V_4	V_L
Silicion					
Germanium					

4 Repeat using the silicon diodes.

Results
1 From your findings make a general statement regarding the approximate forward voltage drop across a silicon and a germanium diode.
2 Think of a possible use for diodes connected in series.
3 Determine the result on the load voltage in the event of the following fault conditions:
   a) D_2 s/c (short-circuit),
   b) D_3 o/c (open-circuit).

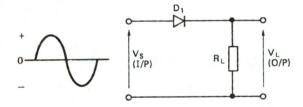

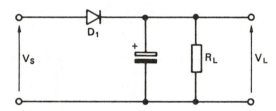

Fig. 1.17 Smoothing circuit

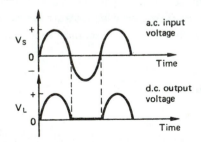

Fig. 1.16 The half-wave rectifier

SMOOTHING

The raw d.c. can be smoothed by connecting a capacitor across the output (Fig. 1.17).

What does the capacitor do?
Connected in this way it acts as a storage vessel or *reservoir* (rather like a tank stores water). It charges up to the peak of the positive half-cycle and then discharges until the next positive peak charges it up again.

You can see from Fig. 1.18 (p. 12) that as the current drawn from the circuit increases, the capacitor discharges more and the output voltage becomes more like the raw d.c.

There is an a.c. component present with the d.c. component. This is called the *ripple voltage* and has the same frequency as the raw d.c. The ideal

PRACTICAL INVESTIGATION 5

Diode clipping circuits

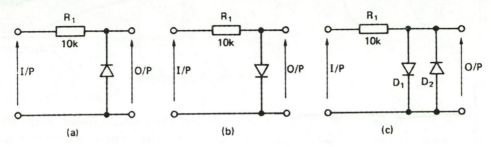

(a)   (b)   (c)

Equipment

Dual beam CRO
Signal generator
4k7 resistor

2 silicon signal diodes
d.c. power supply

Method

1 Build circuit (a).
2 Connect the signal generator to the input and adjust for a 1kHz 8.0V peak-to-peak sine wave (monitor this using one beam of the CRO).
3 Sketch the input (I/P) and output (O/P) waveforms indicating the 0V level and the clipping level.
4 Repeat for circuits (b) and (c).
5 For circuit (c) explain what happens under the following fault conditions:
   a) D_1 o/c,   **b)** D_2 s/c,   **c)** R_1 o/c.

6 Build the circuit below.

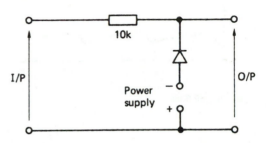

7 With an input of 1kHz 10V peak-to-peak sine wave, adjust the power supply to give negative clipping levels of:
   a) 2.0V,   **b)** 4.0V.
8 Suggest a way in which a specified clipping level could be achieved without a variable power supply.

---

situation is to have zero ripple voltage, but in practice there will always be some ripple present (Fig. 1.19).

REGULATION

This is the ability of a power supply to maintain the output voltage constant despite variations in output current.

Very good quality power supplies have excellent regulation and low ripple (Fig. 1.20) but these require much more circuitry than the simple rectifier circuit so far considered.

CHOICE OF DIODE FOR HALF-WAVE RECTIFIER WITH SMOOTHING

When considering a diode for a half-wave rectifier without smoothing, the maximum reverse bias voltage it will undergo will be the peak of the supply.

i.e. if $V_s = 10V_{rms}$
then $V_{peak} = 10 \times 1.414 = 14.14V$

So, the V_{RRM} of the diode must be greater than 14.14V.

When a smoothing capacitor is used, it charges to V_{peak} during the positive half-cycle. During the

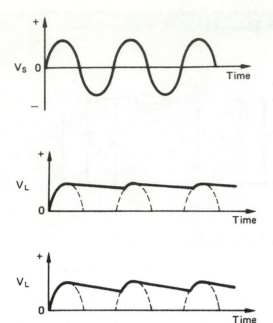

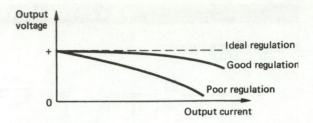

Fig. 1.20 Regulation

negative half-cycle the stored charge reverse biases the diode. The diode now has the negative V_{peak} applied so it now undergoes a reverse bias voltage of $2 \times V_{peak}$ (Fig. 1.21). If $V_s = 10V$, the V_{RRM} of the diode must be greater than 28.28V.

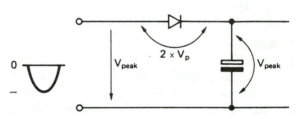

Fig. 1.21 Diode voltage

The disadvantage of half-wave rectification is that there is no diode output for the negative half-cycles. A full-wave rectifier (Fig. 1.22) converts the negative half-cycles into positive half-cycles.

The full-wave bridge rectifier

Operation

During the positive half-cycle, A is positive with respect to B. So, D_1 is forward biased and current flows through D_1, R_L and then D_3 to B (Fig. 1.23).

During the negative half-cycle, A is negative with respect to B. So, B is positive with respect to A! Therefore, D_2 is forward biased, so current flows through D_2, R_L and then D_4 to A (Fig. 1.24).

Choice of diode for full-wave rectification

As you can see from Fig. 1.25, during both half-cycles of the supply voltage two diodes are always conducting. This means that each diode is only

Fig. 1.18 Capacitor action

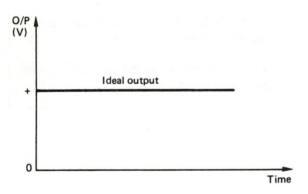

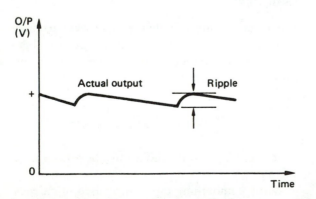

Fig. 1.19 Capacitor output

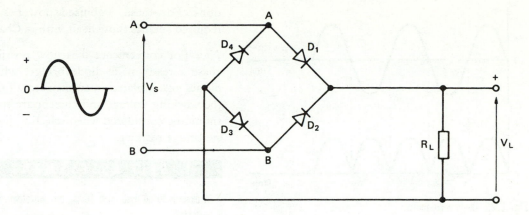

Fig. 1.22 Full-wave bridge rectifier

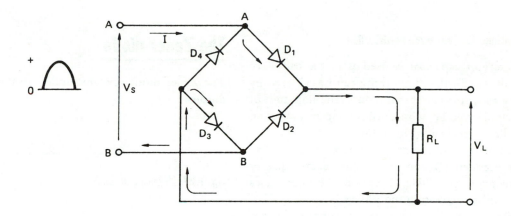

Fig. 1.23 Bridge rectifier operation – positive half-cycle

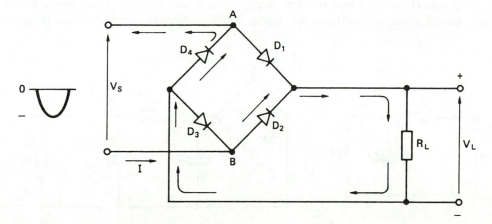

Fig. 1.24 Bridge rectifier operation – negative half-cycle

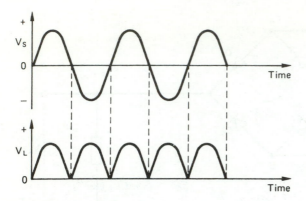

Fig. 1.25 Diode wave forms

subject to a maximum reverse bias voltage of V_{peak} and *not* $2 \times V_{peak}$ as in the half-wave circuit.

Smoothing for full-wave rectification

A reservoir capacitor is used as in the half-wave circuit, but further smoothing can be achieved by using a low-pass filter that reduces the amplitude of the ripple voltage, whilst not significantly affecting the d.c. component (Fig. 1.26).

A low-pass filter is required because d.c. = 0Hz (low frequency!) and the ripple is at a higher frequency, e.g. 100Hz. Therefore the filter passes d.c. whilst attenuating the 100Hz.

These circuits are simple power supplies that can supply only low currents and have relatively poor regulation with quite high ripple voltages. Although they are satisfactory for some applications, e.g. battery chargers and running d.c. motors as in model railways and slot car racing, they are unsuitable for many electronic applica-

tions. For these, stabilised power supplies are required such as those dealt with in Chapter 4.

Note For convenience it is now possible to purchase a ready-made diode bridge, which is four diodes encapsulated in an insulator. Like a diode, the working voltage and current are specified and should be considered when selecting the appropriate bridge rectifier.

The Zener diode

The Zener diode operates under breakdown conditions!

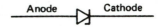

Fig. 1.27 BS Zener diode symbol

Once an ordinary diode has suffered reverse breakdown it is useless. A Zener diode is designed to breakdown when the reverse bias voltage reaches a specified level and current then flows, if the voltage is reduced below breakdown, current ceases to flow. Consider an 8.2V Zener (Fig. 1.28):

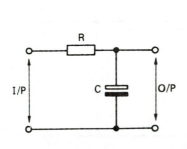

Fig. 1.26 Low-pass filter

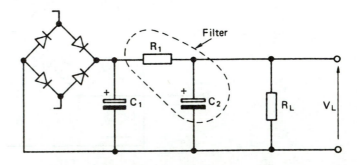

PRACTICAL INVESTIGATION 6

Half-wave rectification

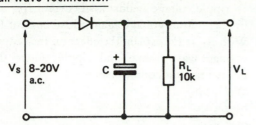

Equipment
Dual beam CRO
Low voltage a.c. source
e.g. transformer 8–20V output
Silicon power diode
Smoothing capacitor 10–33µF
Resistors 10k, 4k7, 1k

Method
1 Connect the circuit as shown above but without capacitor C.
2 Monitor V_S and V_L with the CRO.
3 On a common time scale sketch V_S and V_L, indicating the 0V level and the peak voltages.
4 Connect C and sketch V_L. Measure and record the ripple voltage.
5 Repeat for R_L values of 4k7 and 1k.

Results
1 Under what conditions will the ripple voltage be zero?
2 Suggest a method by which the ripple voltage can be reduced for, say, a load of 1k.

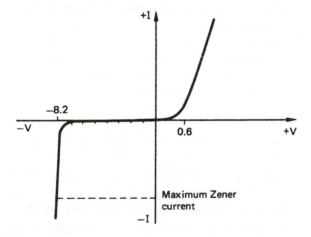

Fig. 1.28 Zener diode characteristics for an 8.2V Zener

The breakdown characteristics for the Zener are much sharper than the normal diode. You can see that in the forward direction it is the same as an ordinary diode.

THOUGHT

■ So a 10V Zener will breakdown at 10V and a 25V Zener at 25V? Yes, but like many components Zener diodes are available in preferred ranges. See the data sheet.

ZENER DIODE SPECIFICATIONS

If you look at the data sheet for Zener diodes you will see that only two specifications are quoted:

1 the nominal Zener voltage (V_Z),
2 the maximum Zener power (P_Z (max)).

Therefore if we want to know the Zener current it must be calculated.

Calculation of Zener current

Since power = voltage × current

$$P = V \times I$$

Zener power = Zener voltage × Zener current

$$P_Z = V_Z \times I_Z$$
$$\therefore I_Z = \frac{P_Z}{V_Z}$$

If P_Z (max) is specified,

$$I_Z \, (\text{max}) = \frac{P_Z \, (\text{max})}{V_Z}$$

The full-wave rectifier

Object

You are required to build a full-wave rectifier with smoothing.

Carry out a full test on the completed circuit and then write a report on your findings.

Equipment

CRO

Mains step-down transformer 240–12V

Four silicon power diodes

22 μF capacitor

10 μF capacitor

Resistors: 270R, 820R, 1k, 2k2, 3k3, 4k7, 6k8, 8k2, 10k

Multimeter

Method

1 Build the rectifier circuit as shown using a load resistor of 10k.

2 Measure and sketch and fully label the a.c. input voltage waveform V_S.

3 Open links (A) and (B). Sketch the waveform V_L indicating 0V and peak values, periodic time and frequency.

4 Reconnect links A and B and make waveform sketches and voltage measurements of V_L, indicating d.c. levels and ripple voltage.

5 Change the value of the load resistor in steps from 10k down to 820Ω. Measure and record the ripple and output voltages and plot a graph of:

a) ripple voltage against load current I_L,

b) d.c. output voltage against load current I_L.

6 With R_L at 1k, explain the effect on the output voltage wave shape and ripple under the following fault conditions:

a) D_1 o/c

b) D_3 s/c

c) C_1 o/c

Test report

This should include the following:

1 Circuit diagram.

2 Brief description of circuit operation.

3 Waveform sketches showing input signal, rectified output, output with smoothing and filtering.

4 Specifications, e.g. output voltage, ripple amplitude, graphs showing variations with load current.

5 Information about the regulation of your power supply based on the output voltage under no load and full load conditions.

(Take full load to be $R_L = 1$k.)

Load regulation =

$$\frac{\text{change in d.c. output voltage}}{\text{d.c. output under no load condition}} \times 100\%$$

6 Basic fault diagnosis information indicating likely faults and their symptoms.

7 From your results present a brief overview of the performance including possible application and limitations.

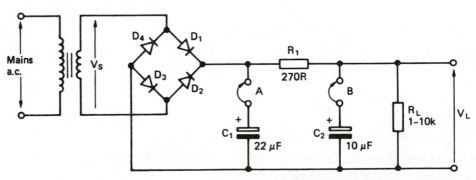

PRACTICAL INVESTIGATION 7

Zener diode operation

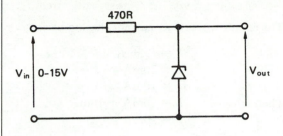

Equipment
d.c. variable power supply
d.c. voltmeter
470R resistor
5.6V Zener diode (400mW)
8.2V Zener diode (400mW)

Method
1 Build the circuit above using a 5.6V Zener diode.
2 Monitor V_{in} and V_{out} and record V_{out} for values of V_{in} from 0 – 15V at intervals of 0.5V.
3 Repeat the above for the 8.2V Zener diode.

Results
1 Plot graphs of output voltage against input voltage for both diodes (on the same axis).
2 From your results state for each diode the change in output voltage that occurs over the input voltage range
   a) 5 – 8V
   b) 9 – 12V.
3 State the maximum current that can be allowed to flow through each diode.

e.g. Consider a BZY88.C20 (20V 400mW Zener diode)

$$I_Z (\text{max}) = \frac{400 \times 10^{-3}}{20}$$
$$= 20 \times 10^{-3} = 20\text{mA}$$

So the maximum reverse bias current that can be passed through this device is 20mA, any more and the device will be destroyed.

In order to avoid the 'knee' at breakdown, the minimum Zener current must be avoided. Usually, however, I_Z (min) is low enough to be ignored, and for calculation purposes the diode may be considered to be ideal, i.e.

$$I_Z \text{ (min)} = 0.$$

Now carry out the practical investigation into Zener diode operation.

■ *There is a maximum Zener current but is there a minimum current? Yes. Let's look closer at the characteristic at breakdown (Fig. 1.29).*

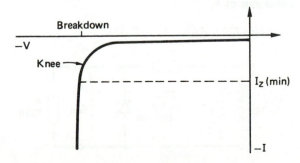

Fig. 1.29 *Zener breakdown characteristics*

USES FOR THE ZENER DIODE

From your investigation you can see that a Zener diode always has its nominal specified voltage across it, i.e. a 10V Zener diode will have 10V across it once breakdown has occurred. This characteristic means that this type of diode can be used for the following applications.

Overvoltage protection

If the supply voltage exceeds 12V the Zener will break down, excessive current will flow, blowing the fuse.

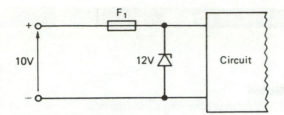

Fig. 1.30 Overvoltage protection

Limiter or clipping circuits

The Zener will break down at 5.6V, preventing the output rising much above this value. (See Practical Investigation 8.)

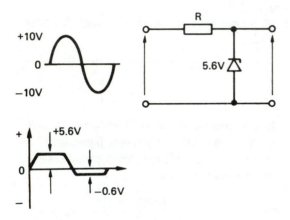

Fig. 1.31 Limiter or clipping circuit

Voltage stabilisation

This is perhaps the most common use for the Zener diode and is based upon the fact that the voltage across the diode remains almost constant for wide variations in input voltage once breakdown has occurred.

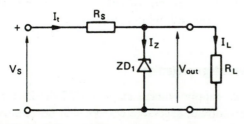

Fig. 1.32 Simple Zener stabiliser

V_{out} will always be the Zener voltage,

i.e. if
$$ZD_1 = 8.2V$$
$$V_{out} = 8.2V$$
$$\therefore V_{out} = V_Z$$

The circuit works on a current sharing basis

$$I_t = I_Z + I_L$$
So, if $\quad\quad I_Z = 20\text{mA}$
$$I_L = 10\text{mA}$$
Then $\quad\quad I_t = 20\text{mA} + 10\text{mA}$
$$= 30\text{mA}$$

If the load were disconnected, how much current would flow through the Zener?

$$I_Z = I_t - I_L$$
$$I_t = 30\text{mA}$$
$$I_L = 0A$$
$$\therefore I_Z = 30 - 0$$
$$= 30\text{mA}$$
So, $\quad\quad I_t = I_Z$ when $I_L = 0$
similarly $\quad\quad I_t = I_L$ when $I_Z = 0$

THOUGHT

- If V_s is 12V and ZD_l is 10V, where does the 2.0V go?
  The voltage ($V_S - V_Z$) is dropped across resistor R_S.
- What happens if V_S goes up?
  V_Z remains constant so the voltage dropped across R_S goes up, i.e. if V_S increases to 14V, V_Z is still 10V.

$$V_S - V_Z = 14 - 10$$
$$= 4V$$

Evidently the value of R_S is important because it limits the current flowing in the circuit.

Calculation of R_S

Consider the circuit shown in Fig. 1.33.

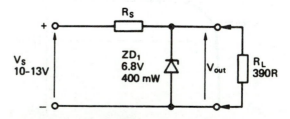

Fig. 1.33 Circuit for the calculation of R_S

PRACTICAL INVESTIGATION 8

Zener diode limiter

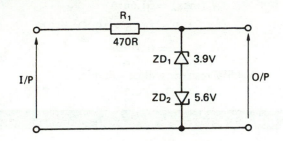

Method
1 Build the above circuit.
2 Connect the signal generator and adjust for a 1kHz sine wave 40V peak-to-peak.
3 Monitor the input and output waveforms with the CRO.
4 Sketch on a common time scale the input and output waveforms indicating the 0V level and clipping levels.
5 Explain what happens under the following fault conditions:
   a) ZD_1 o/c, **b)** ZD_1 s/c, **c)** ZD_2 o/c, **d)** ZD_2 s/c.

Equipment
Dual beam CRO   5.6V Zener diode (400mW)
Signal generator   3.9V Zener diode (400mW)

---

1 Determine load current I_L:

$$I_L = \frac{V_{out}}{R_L}$$

and
$$V_{out} = V_Z$$

$$\therefore I_L = \frac{V_Z}{R_L} = \frac{6.8}{390}$$

$$= 17.4mA$$

2 The load current I_L can be assumed to be constant because R_L is fixed.

3 What is the maximum Zener current I_Z (max)?

$$\frac{P_Z\,(max)}{V_Z} = \frac{400 \times 10^{-3}}{6.8} = 58.8mA$$

4 When will 58.8mA flow through the Zener diode? Under no load conditions and then

$$I_t = I_Z$$

5 R_S *must* limit the current so that with R_L disconnected it will be no greater than 58.8mA.

6 $R_S = \dfrac{V_S\,(max) - V_Z}{I_Z\,(max)}$

$$= \frac{13 - 6.8}{58.8mA} = 105.4\Omega$$

7 A nearest preferred value of resistor to 105.4Ω is 120Ω (120R).

8 *Quick check required!*
   Using the resistor value calculate the Zener current under V_S (max) conditions:

$$V_S\,(max) = 13V$$

$$I_Z = \frac{V_S\,(max) - V_Z}{R_S}$$

$$= \frac{13 - 6.8}{120}$$

$$= \frac{6.2}{120} = 51.6mA$$

9 Under normal operating conditions what is the Zener current when:

   a) $V_S = 10V$,
   b) $V_S = 13V$?

   Note I_L is constant at 17.4mA.

   a) When $V_S = 10V$

$$I_t = \frac{V_S - V_Z}{R_S}$$

$$= \frac{10 - 6.8}{120} = 26.6mA$$

$$I_Z = I_t - I_L$$

$$= 26.6mA - 17.4mA$$

$$= 9.2mA$$

$$I_Z = 9.2mA$$

9 b) When $V_S = 13V$

$$I_t = \frac{V_S - V_Z}{R_S}$$

$$= \frac{13 - 6.8}{120} = 51.6\text{mA}$$

$$I_Z = 51.6\text{mA} - 17.4\text{mA}$$

$$= 34.2\text{mA}$$

10 What is the minimum power rating for R_S?

$$P = I^2R$$

$$R = 120\Omega$$

$$I_t\,(\text{max}) = 51.6\text{mA}$$

$$\therefore P = (51.6 \times 10^{-3})^2 \times 120$$

$$= 0.319\text{W}$$

A 0.5W resistor will be suitable.

DESIGN ASSIGNMENT 2

Zener diode

You wish to operate a small cassette tape recorder from the cigar lighter socket on your car dashboard. The recorder requires a 7.5V d.c. supply and has a 100mA normal operating current. The car battery voltage is nominally 12V, but when travelling at speed the alternator is charging the battery and the terminal voltage could be as high as 14.6V. When stopped at traffic lights, the terminal voltage could be as low as 11V (depending on what is operating; lights, wipers, heated rear screen etc).

Design a simple stabiliser circuit that will ensure that at all times the recorder has the specified voltage and current and under no conditions does any component suffer from overload.

Procedure

1 Start by drawing the basic circuit shown in Fig. 1.34

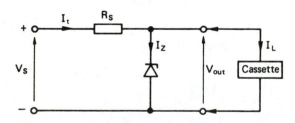

Fig. 1.34 Basic Zener stabiliser circuit

2 State the known facts

$$V_S = 11 - 14.6V$$

$$V_{out} = V_Z = 7.5V$$

$$I_L = 100\text{mA}$$

3 Now for the unknown facts:

I_t

Assuming the diode is ideal ($I_Z\,(\text{min}) = 0$) we can say that

$$I_t = I_L = 100\text{mA}$$

R_S

The function of R_S is to limit the current flowing in the Zener diode under no load conditions.

But remember 100mA must be available for the load under all conditions. That is, when $V_S = 11V$ and $V_S = 14.6V$.

So the worse case for sufficient load current is when V_S is *minimum* at 11.0V.

Calculate R_S for V_S (min)

$$R_S = \frac{V_S\,(\text{min}) - V_Z}{I_t}$$

$$= \frac{11.0 - 7.5}{100 \times 10^{-3}} = 35\Omega$$

The nearest preferred value would be 33Ω

Now check to see what current will actually flow if 33Ω is used.

$$I_t = \frac{V_S(\text{min}) - V_Z}{R_S}$$

$$= \frac{11.0 - 7.5}{33} = 106\text{mA}$$

Because this is higher than 100mA, 33Ω will be suitable.

$$R_S = 33\Omega$$

What current will flow when V_S is maximum at 14.6V?

$$I_t = \frac{V_S(\text{max}) - V_Z}{33}$$

$$= \frac{14.6 - 7.5}{33} = \frac{7.1}{33} = 215.1\text{mA}$$

I_Z
Under the 'worst case' situation V_S will be at a maximum and the load will be disconnected.

$$I_Z = I_t = 215.1\text{mA}$$

So, the Zener power (P_Z) will be

$$I_Z \times V_Z = 215.1 \times 10^{-3} \times 7.5$$

$$= 1.61\text{W}$$

Component selection

Zener diode
A 7.5V Zener diode is required with a maximum power rating in excess of 1.61W. Refer to the data sheet and choose one.
BZX70C7V5 (7.5V 2.5W).

R_S power rating
Under worst case conditions 215.1mA flows.

$$P = I^2R$$

$$= (215.1 \times 10^{-3})^2 \times 33$$

$$= 1.526\text{W}$$

Therefore a 33Ω resistor of 2W or greater is required.

THOUGHT

■ Can a Zener diode keep the voltage constant when the supply V_S is constant but the load current I_L is varying? Yes it can. Remember we said that $I_t = I_Z + I_L$. So, if I_L = goes down I_Z goes up, and if I_L goes up I_Z goes down.

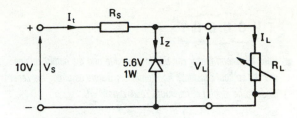

Fig. 1.35 Stabiliser with varying I_L

When R_L is maximum, i.e. infinity (open circuit), $I_L = 0$ (see Fig. 1.35)
Under these conditions

$$I_Z = I_t$$

and

$$I_t = I_Z(\text{max})$$

$$\therefore I_Z = \frac{1W}{5.6V} = 178.6\text{mA}$$

$$R_S = \frac{V_S - V_Z}{I_Z(\text{max})} = \frac{10 - 5.6}{178.6 \times 10^{-3}}$$

$$= 24.6\Omega$$

The nearest preferred value is 27Ω.
With $R_S = 27\Omega$

$$I_Z = \frac{10 - 5.6}{27} = 163\text{mA}$$

The equation $I_t = I_Z + I_L$ must be true for all values of I_L if V_{out} is to remain constant.

$$I_Z + I_L = 163\text{mA} = I_t$$

if $I_L = 50$mA under normal operating conditions,

$$I_t = 163\text{mA} = I_Z + I_L$$

so
$$I_Z = 163\text{mA} - 50\text{mA}$$

$$= 113\text{mA}$$

If I_L goes up to 60mA

$$I_Z = 163 - 60 = 103\text{mA}$$

If I_L goes down to 40mA

$$I_Z = 163 - 40 = 123\text{mA}$$

At all times I_t remains the same

THOUGHT

■ Does this mean that I_L can rise to any value and the output voltage will still be kept constant? No. Remember we are assuming the Zener diode to be ideal, i.e. minimum Zener current = 0.

When

$$I_Z = 0$$

$$I_t = 163mA$$

$$\therefore I_L = I_t - I_Z = 163mA$$

So when I_L is 163mA this is the maximum load current.

Calculation of R_L (min)

$$I_L = 163\text{mA} = I_t$$

$$V_{\text{out}} = 5.6\text{V}$$

$$\therefore R_L(\text{min}) = \frac{V_{\text{out}}}{I_t} = \frac{5.6}{163 \times 10^{-3}} = 34\Omega$$

SELF ASSESSMENT 4

1 From the data sheets, determine:
   a) the maximum forward current of BAV10,
   b) the PIV of 1N916.
2 What is the maximum Zener current for:
   a) BZX85C6V8
   b) BZY93C36
3 A 15V Zener diode rated at 1W is used to stabilise a d.c. voltage of 25V that varies by ± 20%. Calculate the value of the series resistor R_S required.
4 Sketch the forward and reverse characteristics of a 20V, 400mW Zener diode. Indicate the maximum Zener current.

Diode review

■ A diode is a one way 'valve' for current flow.
■ The BS symbol indicates the direction of conventional current flow through the diode (Fig. 1.36).

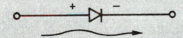

Fig. 1.36 BS diode symbol

■ The two common semiconductor diode materials are silicon and germanium.
■ A diode does not pass any appreciable current unless it is forward biased by making the anode more positive than the cathode, by 0.6V in the case of silicon and 0.3V in the case of germanium.
■ A diode has a very high resistance when reverse biased and a low resistance when forward biased.
■ The following conditions will destroy a diode:

   a) if it is subject to a higher reverse bias voltage than its specified V_{RRM} (max),
   b) if a higher forward bias current is passed through it than that which is specified,
   c) if it is allowed to overheat, i.e. by failing to use a heat sink where appropriate.

■ A very common application for diodes is as a rectifier.
■ The Zener diode operates under breakdown conditions.
■ A Zener diode is connected in reverse bias mode (Fig. 1.37).

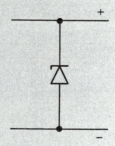

Fig. 1.37 Zener diode

■ Zener diodes can be used for the following applications.
   Overvoltage protection.
   Limiting or clipping circuits.
   Voltage reference sources.
   Simple voltage stabilisation circuits.

SELF ASSESSMENT ANSWERS

Self Assessment 1

1 Negative charge.
2 In the outermost shell.
3 A conductor (very low resistance).

Self Assessment 2

1 Four valence electrons.
2 Yes.
3 Electron hole pairs are generated thermally in pure Ge. If there are 3×10^{28} electrons there will be 3×10^{28} holes.

4

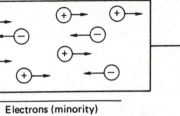

Holes (majority)

Electrons (minority)

(a)

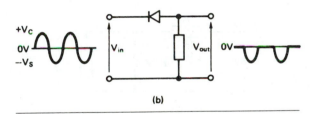

(b)

Self Assessment 3

1 40V r.m.s. has a peak of $40 \times 1.414 = 56.56V$. The PAD5 has a V_{RRM} (max) of only 45V, therefore it is *not* suitable.
2 Germanium has a relatively low temperature capability (80°C) and because power diodes pass high currents leading to the generation of heat the cooling of germanium becomes a problem, so silicon is chiefly used.

Self Assessment 4

1 a) BAV10; $I_F = 200mA$.
   b) 1N916; PIV $= V_{RRM}$ (max) $= 100V$.
2 a) BZX85C6V8 is a 6.8V, 1.3W Zener diode.

$$I_Z \text{ (max)} = \frac{1.3}{6.8} = 191mA$$

   b) BZY93C36 is a 36V, 20W Zener diode.

$$I_Z \text{ (max)} = \frac{20}{36} = 555mA$$

3 20% of 25V = 5V. V_S (min) $= 25 - 5 = 20V$

$$V_S \text{ (max)} = 25 + 5 = 30V$$

$$\therefore \text{ maximum Zener current, } I_Z \text{ (max)} = \frac{1}{15}$$
$$= 66.6mA$$
$$R_S = \frac{V_S \text{ (max)} - V_Z}{I_Z \text{ (max)}} = \frac{30 - 15}{66.6 \times 10^{-3}} = 225.2\Omega$$
nearest preferred value (n.p.v.) $= 270\Omega$

4

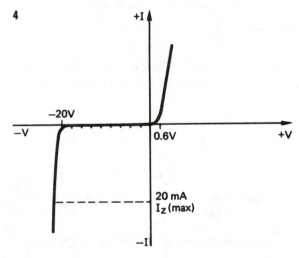

Multiple choice questions

1 The ripple frequency of a full-wave rectifier would be?
a) 50Hz
b) 100Hz
c) 25Hz
d) 200Hz

2 The output voltage (V_{out}) of the circuit shown in Fig. 1.38 would be?
a) 1.2V
b) 5.4V
c) 4.8V
d) 0.6V

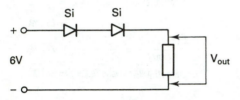

Fig. 1.38

3 Select from the options given in Fig. 1.39 the circuit that indicates a forward biased diode:

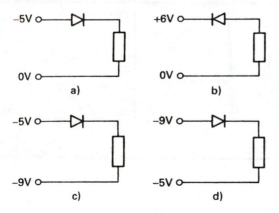

Fig. 1.39

4 A Zener diode coded 1N5342 (see data sheet) is to be used as a voltage stabiliser. What is the maximum current that it will safely pass?

a) 5A
b) 1.36A
c) 735mA
d) 34mA

5 A diode is to be used as a 3A half wave rectifier when connected to a 440V r.m.s. supply. Using the data sheets select the most suitable diode from the options given if a smoothing capacitor is not incorporated.
a) 30S4
b) 1N5626
c) G1756
d) 30S8

6 With reference to Fig. 1.40. The current (I_d) flowing in the Zener diode will be?
a) 98mA
b) 70mA
c) 30mA
d) 100mA

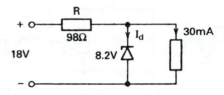

Fig. 1.40

7 The forward voltage drop of a silicon rectifier diode is approximately?
a) 0.2V
b) 0.06V
c) 300mV
d) 600mV

8 The majority carriers in a piece of p-type semi-conductor are?
a) Electrons
b) Protons
c) Holes
d) Negative charge carriers

TRANSISTORS

The transistor was developed in 1948, the name being derived from its original title 'transfer resistor'. In practice, it is a three-terminal device in which current flowing between two terminals can be controlled by a signal on the third terminal. This facility means that a transistor has properties that enable it to be used as an electronic switch and an amplifier.

There are many types of transistor available, but they can effectively be grouped into two families:

- the bipolar junction transistor (BJT)
- the unipolar or field effect transistor (FET)

The bipolar junction transistor

This transistor is a current operated device.

CONSTRUCTION

The bipolar transistor is a three-layer semiconductor device that has three electrodes; the *base*, the *collector* and the *emitter*. There are two types of transistor n-p-n and p-n-p. The construction details together with the BS symbols are shown in Fig. 2.1.

The names of the electrodes come from the parts they play in the operation of the device:

- the Emitter *emits* electrons
- the Collector *collects* the electrons
- the Base *controls* the flow of electrons from emitter to collector

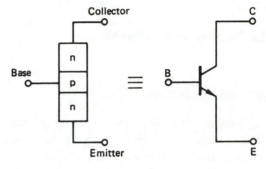

n-p-n transistor

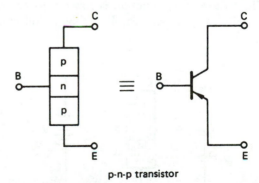

p-n-p transistor

Fig. 2.1 The bipolar transistor

The fact that a transistor is made up of a 'semiconductor sandwich' of p-type and n-type materials means that there are *two* charge carriers involved, hence the name *bipolar*. However, to understand the simple operation of the device it is only necessary to consider one of the charge carriers; electrons for an n-p-n device or holes for a p-n-p device.

For an n-p-n transistor the electrons flow from the emitter region into the base region and continue on to be collected in the collector region. Because conventional current flow is opposite to the electron flow, the arrow on the circuit symbol indicates the flow of *conventional current* through the device (Fig. 2.2).

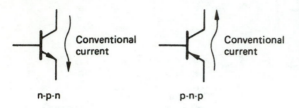

n-p-n	p-n-p

Fig. 2.2 Conventional current flow through transistors

BIASING THE TRANSISTOR

Just as a diode has to be correctly biased in order for current to flow through it, so does a transistor. Remember, the term *biasing* means applying a potential difference (voltage) that establishes the flow of current (Fig. 2.3).

The base–emitter junction is *forward* biased by V_1 whilst V_2 *reverse* biases the base–collector junction.

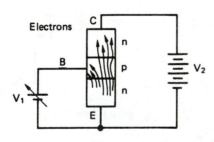

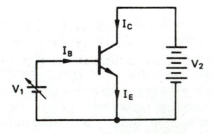

Fig. 2.3 Biased transistor

OPERATION

If V_1 is zero, the current flowing in the base will be zero. Consequently, there will be no current flowing from collector to emitter apart from a small amount of leakage current made up of minority carriers.

If V_1 is now increased, the base–emitter junction becomes forward biased and once the threshold voltage has been reached (0.6 V) base current will flow and then current will flow from collector to emitter.

Note You will notice that the base–emitter voltage (V_{BE}) required to initiate current flow is the same as that required for a forward biased diode. This is because the base–emitter junction of a transistor is in fact a forward biased diode (Fig. 2.4).

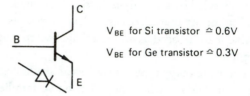

V_{BE} for Si transistor $\simeq 0.6$V

V_{BE} for Ge transistor $\simeq 0.3$V

Fig. 2.4 Base–emitter junction

If the voltage V_1 is further increased, the base current will rise and so will the collector current. The important point to note is that the base current will be small and the collector current much larger, so a *small* current is controlling a *large* current: the device has therefore got amplifying qualities.

BJT notation

The way a transistor operates concerns voltages across, and current flow at, its electrodes. Consequently there needs to be a labelling convention in order to be able to refer sensibly to the device under operational conditions.

Under d.c. or static conditions (no input signal applied) upper case symbols and subscripts are used:

I_B = base current
I_E = emitter current
V_{BE} = voltage between base and emitter
V_{CE} = voltage between collector and emitter

The subscripts here indicate the two points that the potential difference is measured across and quote the most positive electrode first, e.g. for V_{BE} the base is more positive than the emitter (Fig. 2.5). This seems a little complex but, as you will later discover, it becomes very useful when we have to consider the device under a.c. or signal conditions.

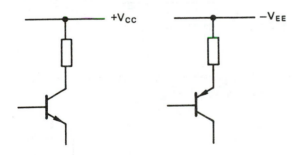

Fig. 2.5 d.c. voltages and currents

If you are familiar with circuit diagrams you may have wondered why the supply rail is often labelled $+V_{CC}$ or $+V_{BB}$. This is in accordance with the British Standards for semiconductor notation: the transistor electrode connected to the supply rail is repeated in the subscript (Fig. 2.6).

Fig. 2.6 Power supply notation

BJT selection

A casual glance at a distributor's catalogue or the specifications included here will quickly indicate that there are literally thousands of transistors available, each one identified by a manufacturer's code.

Like diodes, transistors have voltage, current and power ratings that must not be exceeded or they will be destroyed. The transistor is an amplifying device, so the amount it amplifies, or its *gain*, must be considered along with its operating frequency. A transistor can usually be purchased in p-n-p or n-p-n form and made from silicon or germanium. (Although, like diodes, silicon is the main transistor material, with germanium devices rather scarce today.)

You will also notice that manufacturers indicate suitable uses for their products, e.g. low frequency switch, high frequency amplifiers. This is to help people select an appropriate transistor for their needs.

TRANSISTOR PACKAGE AND PIN CONNECTION

Against each transistor code number is a second code that indicates the package and pin connection for that particular device. This is *most important* because physically many transistors look alike but have a different pin configuration.

Consider a transistor with a T0–92 package and refer to the transistor pin connection sheet. You will see that there are three pin configurations for a straight lead device, and unless you know the package, correct connection is going to be the result of luck rather than knowledge.

A SHORT GLOSSARY OF TRANSISTOR TERMS

Manufacturers and distributors include details in their catalogues that relate to the devices' performance. The list of definitions below will help you to understand and use published specifications.

P_T – The total power that the device can dissipate before damage or destruction occurs.

I_C (max) – The maximum collector current that can be passed without damage.

V_{CB0} – The maximum d.c. voltage between collector and base terminal when the emitter is open circuit.

V_{CE0} – The maximum d.c. voltage between collector and emitter terminal when the base is open circuit.

h_{FE} – The static forward current transfer ratio. (The ratio of d.c. output current to input current – the *current gain*).

F_T – The transition frequency. This is the frequency at which the h_{FE} falls to unity (1).

PRACTICAL INVESTIGATION 9

Transistor operation

Equipment
Power supply
Circuit board
BFY51 transistor
33k, 22k, 10k, resistors
12V, 1 or 2W lamp
Digital multimeter

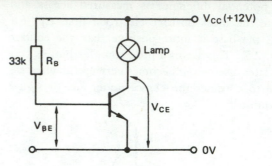

Method
1 Connect the circuit as shown above.
2 Observe that the lamp lights but not brightly.
3 Measure the record voltages V_{BE} and V_{CE}.
4 Repeat the above using the 22k and then the 10k resistor as R_B.

Results
1 Compare the values obtained for V_{BE} and V_{CE} for each of the three base bias resistors.
2 Form a conclusion between the brightness of the lamp and the value of V_{CE}.

From Practical Investigation 9 you should have discovered that when the base–emitter voltage (V_{BE}) is below 0.6V the lamp does not light. This is because the forward conduction voltage has not been reached so no base current and hence no collector current can flow. The higher V_{BE} becomes the brighter becomes the lamp, simply because if V_{BE} is increased, I_B increases and so does I_C. However, you will have found that the maximum V_{BE} that can be made is about 0.7V; the transistor is then fully conducting and the lamp at its brightest.

It is interesting to monitor V_{CE} and find that with the lamp just on, V_{CE} is quite high. The brighter the lamp becomes, the smaller V_{CE} becomes. This is due to the fact that as the collector current increases, the voltage drop across the lamp increases and V_{CE} goes down. For example, if there is a supply of 12V, and 8V is dropped across the lamp, V_{CE} can only be 4V.

Essentially then, this investigation serves to confirm the operation of a transistor and shows that the current flowing from collector to emitter is controlled by the base current.

So far we have considered a transistor connected as shown in Fig. 2.7.

This is the *common emitter configuration* (or *connection* or *mode*). It is so called because the emitter electrode is common to both the input and output

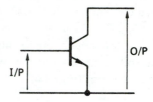

Fig. 2.7 Common emitter configuration

signal. It is possible to connect the device in two other modes (Fig. 2.8).

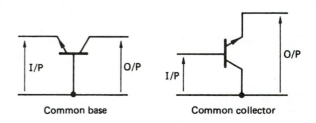

Fig. 2.8 Alternative configurations

Each connection offers advantages over the other two that makes it suitable for certain applications, but because the common emitter configuration is the most frequently used we shall only consider this circuit.

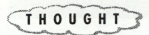

- *What about a p-n-p transistor?*
  The operation is similar, but the circuit symbol tells us that the current flows the opposite way (Fig. 2.9).

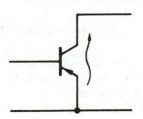

Fig. 2.9 p-n-p transistor operation

For this to happen the transistor must be biased differently (Fig. 2.10).

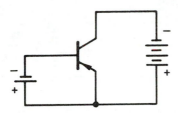

Fig. 2.10 p-n-p transistor bias

The base–emitter junction is still forward biased and the base collector junction is reverse biased, just as for the n-p-n, but to achieve this the polarities of the bias voltages are reversed. Other than this it performs identically.

Note V_{BE} *is* $-0.6V$ *for a p-n-p transistor. Therefore* I_B *flows the opposite way to an n-p-n device.*

BJT characteristics

The characteristics of a diode showed how it performed under operating conditions. The transistor also has characteristics in the form of graphs or curves that are useful in predicting its performance, but because it is a control device, a set of characteristics must be drawn for the input and output conditions to show how it operates over a range of values.

INPUT CHARACTERISTICS

This shows how the input current varies with input voltage (Fig. 2.11).

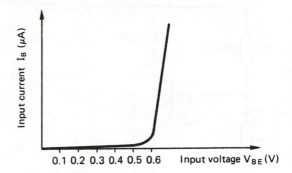

Fig. 2.11 Input characteristics

Note This is similar to the forward biased diode.

OUTPUT CHARACTERISTICS

This shows how the output current varies with the output voltage for a given fixed value of input current (Fig. 2.12). Because the input current determines the output current, it is possible to draw an infinite number of curves on this graph. So it is usual for manufacturers to show a limited number of curves over the operating range.

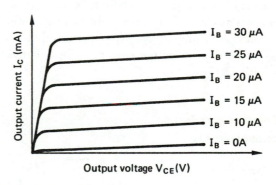

Fig. 2.12 Output characteristics

Note When the input current is zero, the output current should be zero, but in reality there is a minute amount of leakage current due to minority carriers. However in modern silicon transistors this is very low – just as it was for a silicon diode. The next investigation will enable you to plot your own characteristics.

PRACTICAL INVESTIGATION 10

Transistor performance characteristics

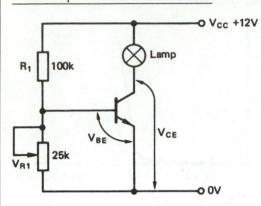

Equipment
Power supply
breadboard
10k resistor
12V lamp (1–2W)
n-p-n transistor (I_c = 150mW max.) e.g. BFY 51
Three multimeters

Method
1 Connect the circuit as shown. Monitor I_B using a 0–0.5mA meter and V_{BE} using a 0–1V meter.
2 Adjust the power supply to +12V.
3 By varying V_{R_1} measure V_{BE} for steps in I_B from 0.05mA to 0.5mA.
4 Switch off the supply and reconnect using meters to monitor I_B and I_C.
5 Set the supply to 5V and adjust V_{R_1} to give I_B = 0.1mA.
6 Monitor and record I_C for values of V_{CE} from 5V to 12V, *but* keep I_B constant using V_{R_1} for each step change.
7 Repeat the above for values of I_B of 0.2mA and 0.3mA.

Results
1 Plot the graph of I_B against V_{BE} (input characteristics).
2 Plot the graph of I_C against V_{CE} for values of I_B of 0.1mA, 0.2mA and 0.3mA. (Output characteristics.)

TRANSISTOR CURRENT GAIN

As stated in the Glossary, the current gain of a transistor is called h_{FE} when it is in common emitter mode (in common base and common collector it would be h_{FB}, h_{FC} respectively). The h_{FE} is the ratio of output current to input current (see Fig. 2.13).

Output current = I_C

Input current = I_B

$$\therefore h_{FE} = \frac{I_C}{I_B}$$

Fig. 2.13 Transistor current gain

Refer to the data provided for the BC108 and you will notice that the h_{FE} quoted is a range of 110–800 with some manufacturers quoting a typical value of h_{FE} for a specific collector current I_C. Alternatively, only the minimum h_{FE} is quoted

as in the case of a BC184L. The reason for this is that devices are difficult and expensive to make with exact values so it is quite acceptable to indicate the possible extremes of the h_{FE}. The next investigation highlights this phenomenon.

The BJT as an electronic switch

The object here is to use the low base current of a transistor to switch a higher collector current. In this way a transistor can be considered as a switch (Fig. 2.14).

Consider the circuit shown in Fig. 2.15.

When $V_{in} = 0V$

With no base current (I_B) virtually no collector current (I_C) will flow.

$$V_{CE} \simeq V_{CC}$$

So 0V in gives + V_{cc} output

PRACTICAL INVESTIGATION 11

Verification of transistor gain

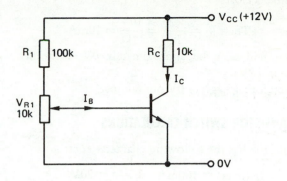

Equipment
Power supply
Circuit board
Three identical transistors
100k, 10k resistors
10k potentiometer
Two digital multimeters.

Method
1 Connect the circuit as shown above.
2 Refer to the data on your chosen transistor and select a value of collector current well below the maximum, e.g. 10 mA.
3 Monitoring I_B and I_C with the meters, adjust V_{R1} until your chosen collector current flows. Record I_B and I_C.
4 Repeat for the other two transistors ensuring that the collector current is always the same.

Result
1 Calculate the h_{FE} for each transistor using

$$h_{FE} = \frac{I_C}{I_B}$$

2 The transistors have the same code number but do they have the same gain?

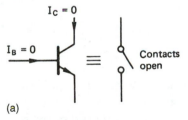

(a)

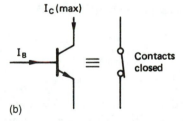

(b)

Fig. 2.14 Transistor switch: (a) open, (b) closed

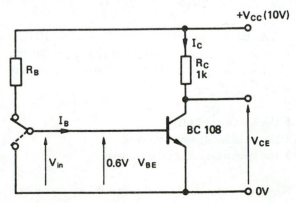

Fig. 2.15 Transistor switch circuit

THOUGHT

0V in gives +10V out?
Yes — using Ohms Law:

$$I_C = 0$$

∴ voltage drop across $R_C = 0V$.

Consequently, if there is no voltage drop across it, both ends will be at the same potential.

When V_{in} is applied

I_B flows so assume that the imaginary switch contacts close.

$$\text{Then } I_C = \frac{V_{CC}}{R_C} = \frac{10V}{1k} = 10mA$$

V_{CE} will now be approximately 0V (in reality about 0.2V).

Therefore V_{in} gives 0V out (almost).

TRANSISTOR SWITCH CALCULATIONS

A BC108 has the following characteristics:

$$I_C \text{ (max)} = 100mA \quad V_{CE0} = 20V$$
$$h_{FE} = 110\text{–}800$$

For the proposed circuit shown in Fig. 2.15,

$$I_C = 10mA$$
$$V_{CE} = 10V \text{ (max)}$$

Therefore the device is well within specification.

Calculation of R_B

When the transistor is off,

$$I_C = 0$$
$$V_{CE} \simeq V_{CC} = +10V \text{ (point A)}$$

When the transistor is on,

$$I_C = 10mA$$
$$V_{CE} \simeq 0V \text{ (point B)}$$

This is called the transistor *load line* (Fig. 2.16) and it shows that the device can be operated anywhere along this line by the use of the appropriate base current.

Because we are using it as a switch, when it is off $I_B = 0$. When it is on I_B must be a value that will turn the transistor fully on in order for 10mA collector current to flow.

Since $h_{FE} = 110\text{–}800$ we assume the lowest h_{FE}:

$$h_{FE} \text{ (min)} = 110$$

$$h_{FE} = \frac{I_C}{I_B}$$

$$I_B = \frac{I_C}{h_{FE}} = \frac{10mA}{110}$$

$$= \frac{10 \times 10^{-3}}{110} = 91\mu A$$

So I_B must be $91\mu A$ (Fig. 2.17).

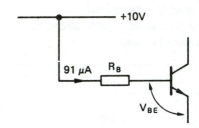

Fig. 2.17 Current through R_B

$91\mu A$ flows through R_B. Now $V_{BE} \simeq 0.6V$ (for the transistor to be turned fully on). So $V_{CC} - V_{BE}$ is dropped across R_B

$$R_B = \frac{V_{CC} - V_{BE}}{91\mu A} = \frac{9.4}{91 \times 10^{-6}} = 103k$$

The nearest preferred value (n.p.v.) to 103k is 100k. So from this you can see that a switch can be designed provided the following facts are known:

1 the load current (I_C)
2 the switching signal voltage (V_{in})
3 the specifications of the transistor.

Now use Practical Investigation 12 to prove this theory.

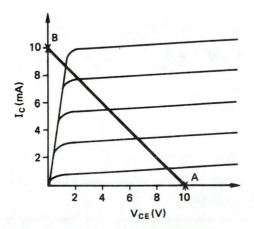

Fig. 2.16 Transistor output characteristics with load line

PRACTICAL INVESTIGATION 12

The BJT as a switch

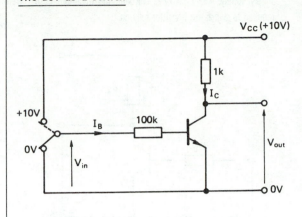

Equipment

Power supply   Circuit board
BC108   Digital multimeter
1k, 100k resistors

Method

1 Connect the circuit as shown and adjust the power supply to +10V.
2 Monitor V_{out} with the voltmeter and record V_{out} when $V_{in} = 0V$ and +10V.
3 By investigation record I_C and I_B when V_{in} is +10V.

Results

1 Does it matter that the transistor may have an h_{FE} in excess of the 110 used in the calculations?
2 Make a general observation about a transistor when used as a switch with regard to V_{in} and V_{out}.

DESIGN ASSIGNMENT 3

Electronic switch

A BJT is to be used as an electronic switch that will enable a 12V, 2W lamp to be controlled by a 5V input voltage capable of delivering no more than 5mA.

Design a circuit that will fulfil the above requirements ensuring that under no conditions is any specification for the selected device exceeded.

Your results must show all calculations and assumptions leading to your choice of transistor and resistor values. Practical verifications of the theoretical results in the form of input and output switching currents and voltages must be included, and a full circuit diagram together with a report on the circuit operation is required.

Procedure

1 Sketch out the basic circuit together with known facts or parameters.

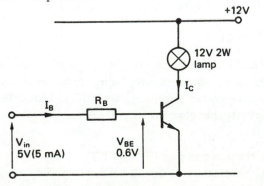

2 Now for unknown facts.

I_C (*max*)

$$I_C = \frac{2W}{12V} = 167mA$$

I_B (*max*)

This must not be greater than 5mA

$$\therefore h_{FE}\,(min) = \frac{I_C}{I_B} = \frac{167mA}{5mA} = 33.4$$

Remember This is the absolute minimum. Anything larger will ensure that the circuit operates comfortably rather than at its limits.

3 Select a transistor that has I_C (max) greater than 167mA, V_{CE0} greater than 12V, h_{FE} (min) greater than 33.4.

I have chosen a BFY 51 (simply because that is what I have readily to hand that meets the specification!).

BFY 51:

$$I_C \text{ (max)} = 1A$$
$$V_{CE0} = 30V$$
$$h_{FE} \text{ (min)} = 40$$

Calculation of R_B

Calculate I_B first, using transistor specification details and the required collector current (I_C)

$$I_C = 167mA$$
$$h_{FE} = 40$$
$$\therefore I_B = \frac{I_C}{h_{FE}} = \frac{167 \times 10^{-3}}{40} = 4.12mA$$
$$V_{in} = 5.0V$$
$$V_{BE} \simeq 0.6V$$
$$R_B = \frac{V_{in} - V_{BE}}{4.12mA} = \frac{4.4}{4.12 \times 10^{-3}} = 1068\Omega$$

The n.p.v. = 1k

HOLD IT! Will the use of 1k exceed the base current of 5mA? Quick check:

if $$R_B = 1k$$
$$I_B = \frac{4.4}{1000} = 4.4mA$$

This is still OK, so the finished design will be:

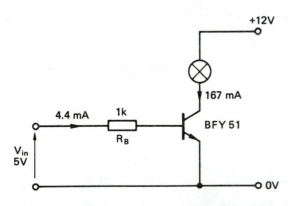

All that is now required is to build, test and report on the design.

1 A BC182L is to be used as a switch controlling a load of 1.5W with a supply voltage (V_{CC}) of 12V. Calculate the minimum base current required to switch the transistor fully on.

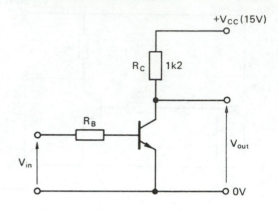

2 a) Sketch the output characteristics of the above transistor and include a load line showing how the two points are determined.

b) Show on the load line the areas that indicate the device is
  i) fully on (saturated), ii) fully off (cut-off).

c) If the transistor has an $h_{FE} = 50$, calculate the value of base resistor R_B if V_{in} has a maximum value of 4V.

3 Sketch the circuit diagram of a p-n-p transistor used to switch a 12V lamp from an input signal of 5V.

The field effect transistor (FET)

This is a voltage controlled device

The FET is sometimes called a unipolar device because the current flow through it consists of only one type of charge carrier, i.e. holes or electrons.

The current flow through a FET is controlled by the voltage across its input terminals (unlike the BJT where current through is controlled by the current at the input).

TYPES OF FET

The FET family is constantly growing as new improved versions are being made that offer greater power handling capabilities, but despite this fast changing situation there are basically two types:

- the junction gate FET (JFET)
- the metal oxide silicon FET (MOSFET)

The JFET

Like the BJT, it is available in complementary form (n or p channel) as shown in Fig. 2.18.

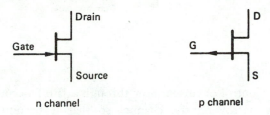

Fig. 2.18 BS symbols for JFETs

The *source* is the source of the electrons, which are collected or drained at the *drain*, whilst the *gate* is the control electrode.

JFET NOTATION

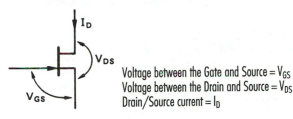

Voltage between the Gate and Source = V_{GS}
Voltage between the Drain and Source = V_{DS}
Drain/Source current = I_D

Fig. 2.19 JFET notation

THOUGHT

■ *What about the gate current I_G?*
This is very low indeed (pA). This factor gives the FET a very high input impedance which is a useful characteristic for many applications.

n CHANNEL JFET

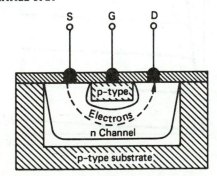

Fig. 2.20 Representation of an n channel JFET

Construction

The channel is a lightly doped, n-type semiconductor and the gate region is a heavily doped, p-type semiconductor.

Between the source and drain is the n channel which is purely resistive, so electrons can easily flow along it.

Biasing

The drain is biased positively with respect to the source by V_{DS}; the gate is biased negatively with respect to the source by V_{GS} (Fig. 2.21).

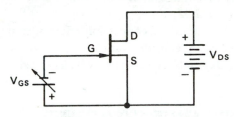

Fig. 2.21 Biasing a JFET

If V_{GS} is zero, then the maximum number of electrons will flow along the channel from source to drain, so maximum current will flow from drain to source (I_D).

If V_{GS} is increased, the gate source voltage becomes negative and less drain current flows. When V_{GS} becomes sufficiently negative, no drain current will flow. Therefore,

when V_{GS} = zero I_D = max (device ON)
when V_{GS} = max I_D = zero (device OFF)

Quick comparison with the BJT:

when I_B = zero I_C = zero (device OFF)
when I_B = max I_C = max (device ON)

It is important to note yet again that it is the input voltage (V_{GS}) that controls the output current (I_D).

p CHANNEL

The situation here is reversed with a p-type channel and an n-type gate region. To bias the device correctly, the drain must be negative and the gate positive with respect to the source.

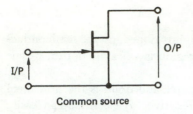
Common source

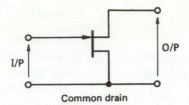
Common drain

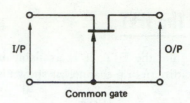
Common gate

Fig. 2.22 JFET configurations

JFET characteristics

Just as the BJT can be connected in three different modes or configurations the same applies to the FET (Fig. 2.22).

Each method of connection offers advantages, but we shall consider only the common source configuration because it is very widely used.

OUTPUT CHARACTERISTICS (n CHANNEL)

This shows how the drain current I_D varies with the drain source voltage for different values of V_{GS} (Fig. 2.23).

You will see that maximum drain current flows when $V_{GS} = 0V$. The point at which the drain current levels off is called the *pinch off* voltage, which is analogous to someone squeezing a hose pipe restricting the bore and limiting the flow.

Control of current flow through a JFET occurs by narrowing the channel so that it becomes depleted of charge carriers. For this reason JFETs as said to operate in *depletion mode*.

THE JFET AS A SWITCH

The FET can be used as an electronic switch where V_{GS} controls I_D and hence V_{out} (Fig. 2.24).

Point A:
when $I_D = 0$, $V_{DS} \simeq V_{DD}$

Point B:
when $I_D = $ max, $V_{DS} = 0V$

$$\therefore I_D \,(\text{max}) = \frac{V_{DD}}{R_D}$$

Now perform Practical Investigation 13 and prove the operation for yourself.

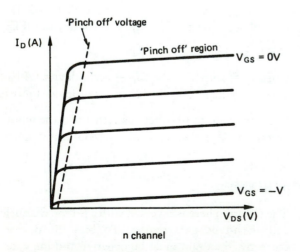
n channel

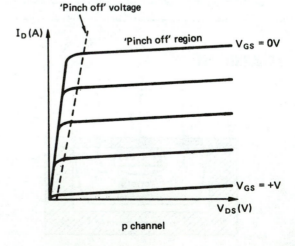
p channel

Fig. 2.23 Output characteristics

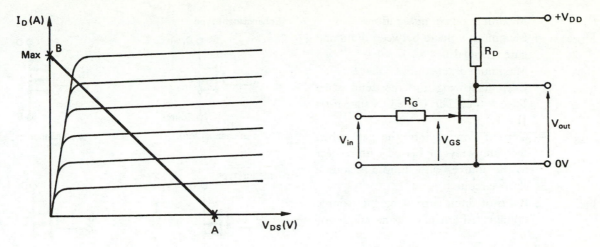

Fig. 2.24 JFET switch output characteristics and circuit

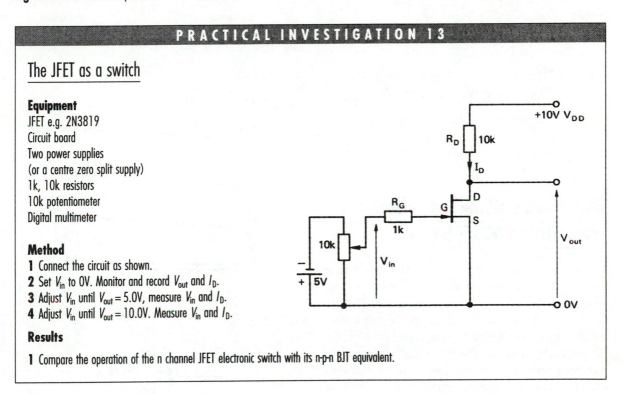

PRACTICAL INVESTIGATION 13

The JFET as a switch

Equipment
JFET e.g. 2N3819
Circuit board
Two power supplies
(or a centre zero split supply)
1k, 10k resistors
10k potentiometer
Digital multimeter

Method
1 Connect the circuit as shown.
2 Set V_{in} to 0V. Monitor and record V_{out} and I_D.
3 Adjust V_{in} until $V_{out} = 5.0V$, measure V_{in} and I_D.
4 Adjust V_{in} until $V_{out} = 10.0V$. Measure V_{in} and I_D.

Results

1 Compare the operation of the n channel JFET electronic switch with its n-p-n BJT equivalent.

Choosing a suitable FET

A field effect transistor has various ratings that must not be exceeded, just like any other component, but they are expressed differently to the BJT. JFETs are available with an n or p channel, and power FETs are of the MOSFET type (yet to be considered). Below are the important parameters that you need to be aware of when selecting a suitable device for a particular application, be it switch or amplifier. This list can be used in conjunction with the FET data provided for reference purposes or any catalogue, but remember manufacturers differ in the way they present the same information. Be prepared for alternative abbreviations to be used.

P_{T} – Maximum power dissipation.

V_{DG} – Maximum voltage between drain and gate terminals.

V_{DS} – Maximum drain-source voltage.

I_{DSS} – Current flowing into the drain when $V_{\mathrm{GS}} = 0V$. (This is the ON state for a JFET.)

I_{GSS} – Current flowing into the gate when the gate is reverse biased with respect to the source, with drain and source short circuited.

Y_{FS} – Ratio of drain current to gate source voltage. Alternative terms are g_{m} or g_{fs}.

I_{D} (max) – Maximum permissible drain current.

Note

Sometimes I_{D} (max) is not given but P_{T} and V_{DS} are:

e.g. BF244A:

$$P_{\mathrm{T}} = 300\mathrm{mW}, \quad V_{\mathrm{DS}} = 30\mathrm{V}$$

$\therefore$ If the maximum V_{DS} for your circuit is 30V:

$$I_{\mathrm{D}}\,(\mathrm{max}) = \frac{300 \times 10^{-3}}{30} = 10\mathrm{mA}$$

Metal oxide silicon field effect-transistors (MOSFETs)

The construction of this type of FET differs from the JFET in that the gate is completely insulated from the channel by a thin layer of metal oxide. This gives an extremely high input resistance (they used to be called IG FETs – insulated gate FETs). The actual physics of the way the device operates has been well documented elsewhere so here only the practical operational characteristics will be considered.

TYPES OF MOSFET

The MOSFET is available in n or p channel as an *Enhancement* or *Depletion* type (Fig. 2.25).

What is the difference between the types?
Let us consider the output characteristic of an n channel enhancement type MOSFET (Fig. 2.26).

With the drain positive with respect to the

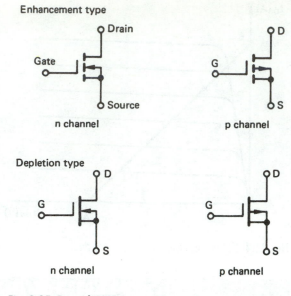

Fig. 2.25 Types of MOSFETs

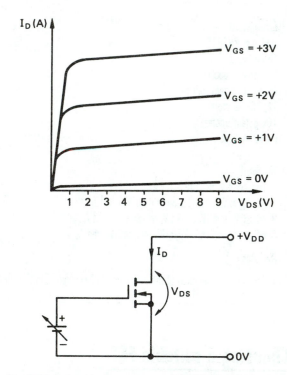

Fig. 2.26 n channel enhancement type MOSFET output characteristics

source and $V_{\mathrm{GS}} = 0$ V virtually no drain current flows. If V_{GS} is made positive, drain current flows. The more positive V_{GS} becomes, the greater the drain current becomes.

Now for the n channel depletion type (Fig. 2.27).

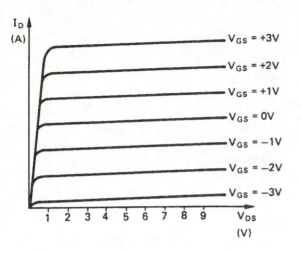

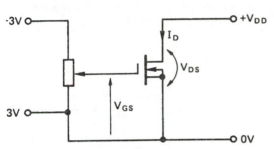

Fig. 2.27 n channel depletion type MOSFET output characteristics

With the drain positive with respect to the source and $V_{GS} = 0V$, drain current flows (just as in the JFET). If V_{GS} is made negative, I_D reduces – just like the JFET. But if V_{GS} is made positive I_D increases. *So the depletion type MOSFET operates in both enhancement and depletion mode.*

MOSFETs are available as power FETs under headings such as DMOS and VMOS. This refers to the way they are made and does not mean that they operate differently. There is also a tendency for power FETs to be available mostly in n channel form as enhancement type devices. Practical Investigation 14 will give you a chance to become familiar with a VMOS FET used as a switch.

SELF ASSESSMENT 6

1 A 2N5460 p channel FET is to be used to control a 250mW 10V lamp switched by a +5V input voltage.
   a) Calculate the drain current that will flow.
   b) Sketch the circuit diagram showing the polarities of the voltages involved.
2 A field effect transistor is required to control the current through a load of 15Ω. If the d.c. supply voltage is 150V, calculate the drain current and select a suitable device from the specifications provided.

Transistor review

- Bipolar junction transistors are current operated devices in which the input current signal controls the output current.
- The V_{BE} of a transistor when fully conducting is similar to the voltage drop across a diode (0.6V for small signal transistors).
- The d.c. current gain of a transistor in common emitter mode is termed the h_{FE}.
- $h_{FE} = \dfrac{I_C}{I_B}$.
- A BJT used as an electronic switch has a very low output voltage when saturated (fully ON) and a high output voltage when cut-off (fully OFF).
- Field effect transistors are voltage operated devices in which the input voltage signal controls the output current.
- Junction FETs operate in depletion mode only. Therefore the input voltage signal can only make the channel narrower.
- For a JFET maximum output current flows when the input voltage signal is zero (I_D (max) when $V_{GS} = 0$).

- MOSFETs are available as enhancement or depletion types.
- With an enhancement FET, the channel can only be widened (enhanced) by the input voltage signal (not reduced).

  – For an n channel enhancement type I_D is max when $V_{GS} = $ max positive.
  – For a p channel enhancement type I_D is max when $V_{GS} = $ max negative.

- A depletion type MOSFET will operate in either mode, so for an n channel device:

  $V_{GS} = $ negative, channel narrows;
  I_D reduces.

  $V_{GS} = $ positive, channel widens;
  I_D increases.

PRACTICAL INVESTIGATION 14

The VMOS FET as a switch

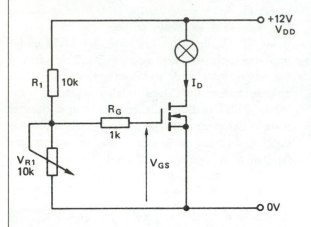

Method

1 Connect the circuit as shown.

2 Adjust V_{R_1} until the lamp *just* lights. Measure and record V_{GS} and I_D.

3 Adjust V_{R_1} until the lamp is at full brilliance. Record V_{GS} and I_D.

4 By investigation, determine the lowest V_{GS} required for I_D to flow.

5 Disconnect the gate and observe what happens to the lamp when the gate electrode is handled.

Equipment
12V 2W lamp
Suitable VMOS FET e.g. VN66AF
Power supply
Circuit board
1k, 10k resistors
10k potentiometer
Digital multimeter

Results

1 Compare the power MOSFET to a conventional power BJT, noting any obvious advantages.

2 From your investigations, what conclusion do you draw relating to the gate terminal and operating conditions?

SELF ASSESSMENT ANSWERS

Self Assessment 5

1 BC182L has $h_{FE} = 100\text{--}480$ $\therefore$ h_{FE} (min) $= 100$.

$$\text{Load current } I_C = \frac{1.5W}{12V} = 125mA$$

$$\text{Minimum } I_B = \frac{125mA}{100} = 1.25mA$$

2

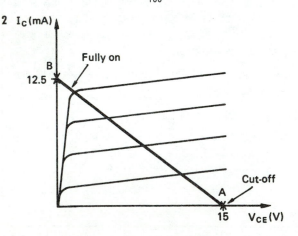

b) Point A $= V_{CE}$ when $I_C = 0$.

$$V_{CE} \simeq V_{CC} = 15V$$

Point B $= I_C$ when $V_{CE} = 0$.

$$I_C = \frac{V_{CC}}{R_C} = \frac{15}{1.2k} = 12.5mA$$

c)
$$I_C = 12.5mA$$
$$h_{FE} = 50$$
$$\therefore I_B = \frac{12.5mA}{50} = 250\mu A$$
$$V_{in} = 4V$$
$$R_B = \frac{V_{in} - V_{BE}}{I_B} = \frac{4 - 0.6}{250 \times 10^{-6}} = 13.6k$$
n.p.v. $= 12k$

3

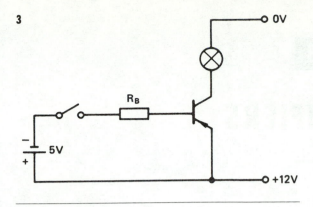

Self Assessment 6

1 a) Lamp power = 250mW
Lamp voltage = 10V

Lamp operating current = $\dfrac{250mW}{10}$ = 25mA

∴ Maximum drain current will be 25mA.

b)

2 If d.c. supply = 150V then V_{DS} (max) = 150V.
If drain resistor = 15Ω

I_D (max) = $\dfrac{150}{15}$ = 10A

∴ Transistor must be capable of V_{DS} (max) = 150V
I_D (max) = 10A

IRF640: V_{DS} = 200V, I_D (max) = 11A.

Multiple choice questions

1 What is the approximate base emitter voltage (V_{BE}) of a silicon transistor?
a) 0.3V
b) 1.2V
c) 600mV
d) 0.06V

2 A transistor with h_{FE} = 150 has a base current (I_B) of 20μA what will the collector current (I_C) be?
a) 300μA
b) 133mA
c) 3mA
d) 0.03mA

3 From the options given choose the one that accurately describes the BJT and FET.
a) The BJT and FET are both current operated devices.
b) The BJT is a current operated device and the FET a voltage operated device.
c) The FET and BJT are both voltage operated.
d) The FET is a current operated device and the BJT a voltage operated device.

4 Consider an n channel JFET connected in common source mode.

a) Maximum drain current (I_D) flows when the gate source voltage (V_{GS}) is zero.
b) Maximum drain current (I_D) flows when the gate source voltage (V_{GS}) is positive.
c) Maximum drain current (I_D) flows when the gate source voltage (V_{GS}) is negative.
d) Maximum drain current (I_D) flows when the gate source voltage (V_{GS}) is 0.6V.

5 With reference to Fig. 2.28. The output voltage (V_{out}) will be?
a) 0V
b) 1.0V
c) 11.0V
d) 0.11V

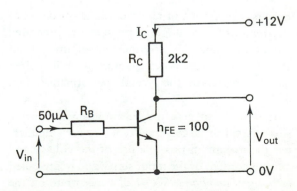

Fig. 2.28

3

AMPLIFIERS

—

The function of an amplifier is to increase the voltage, current or power content of an electrical signal.

There are three types of amplifier and each will provide amplification to any input signal so that the output signal will be greater. This characteristic is termed the *gain* of the amplifier.

Gain is a measure of how much an amplifier amplifies. It is calculated using the input and output signal levels.

$$\text{Gain} = \frac{\text{Output signal}}{\text{Input signal}}$$

Note   Gain is a ratio. Therefore it has **no units**.

The most widely used symbol for gain is A with the subscripts i, v, p denoting the signal involved. Hence:

$$A_i = \text{current gain,}$$
$$A_v = \text{voltage gain,}$$
$$A_p = \text{power gain.}$$

Because the circuit of an amplifier may consist of anything from a single transistor or integrated circuit to a complex arrangement of many transistors and ICs, the block diagram symbol is often used to represent the overall performance (Fig. 3.1).

It is usual to describe an amplifier as either a small signal or large signal type. Power amplifiers use the maximum possible use of the voltage and current content of the input signal and so are called *large signal amplifiers*. We shall concentrate on the small signal amplifier.

Current gain A_i

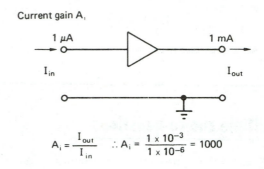

$$A_i = \frac{I_{out}}{I_{in}} \quad \therefore A_i = \frac{1 \times 10^{-3}}{1 \times 10^{-6}} = 1000$$

Voltage gain A_v

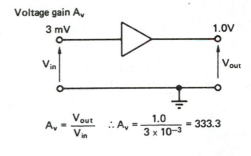

$$A_v = \frac{V_{out}}{V_{in}} \quad \therefore A_v = \frac{1.0}{3 \times 10^{-3}} = 333.3$$

Power gain A_p

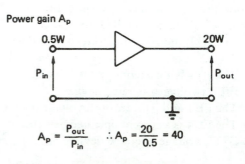

$$A_p = \frac{P_{out}}{P_{in}} \quad \therefore A_p = \frac{20}{0.5} = 40$$

Fig. 3.1 Current, voltage and power gain

The BJT as a small signal amplifier

From your investigations with the BJT as a switch you are familiar with its amplification qualities, namely its h_{FE} which is the forward d.c. current gain of the device itself.

When used as a switch, the input signal simply drives the device from a non-conducting state into a fully conducting, or saturated state, i.e. from OFF to ON.

For the transistor to be employed as a signal amplifier, use must be made of the fact that the transistor can be biased so that it is partly on. Any input signal that is then applied will make the device conduct more or less, with corresponding changes in the voltages and currents involved. Consider the output characteristics and load line shown in Fig. 3.2.

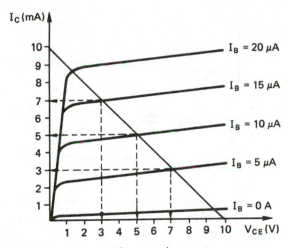

Fig. 3.2 Transistor as amplifier output characteristics

By selecting a suitable d.c. base current (I_B) the transistor can be biased anywhere along the load line

e.g. if $I_B = 0A$, $I_C \simeq 0$, the transistor is OFF
$\therefore V_{CE} \simeq 10V$

if $I_B = 5\mu A$, $I_C = 3mA$, the transistor is partly ON.
$\therefore V_{CE} = 7.0V$

if $I_B = 10\mu A$, $I_C = 5mA$, the transistor is partly ON.
$\therefore V_{CE} = 5.0V$

if $I_B = 15\mu A$, $I_C = 7.0mA$, the transistor is partly ON.
$\therefore V_{CE} = 3.0V$

As you can see the amount of output current flowing in the device is controlled by the choice of input current I_B. This bias point is under d.c. conditions: that means without any signal connected to the input terminals. The correct term for this is *quiescence*. Therefore quiescence means under no signal conditions (d.c. only). Please do not get confused with the d.c. power supply: when we refer to signal conditions it means with an input signal applied to the amplifier. The power supply is there simply to bias the device.

From previous work it should be clear that the quiescent, or Q, point can be anywhere along the load line from the bottom ($V_{CE} = 10V$, $I_C = 0A$) to the top ($V_{CE} = 0V$, $I_C = 10mA$). So how is the position chosen?

An amplifier must provide minimum distortion, i.e. the output must be an exact reproduction of the input only bigger. To achieve this V_{CE} (V_{out}) must be able to swing the maximum possible amount in both directions. Therefore the best position is half-way along the line, such that under quiescent conditions

$$V_{CEQ} = \frac{V_{CC}}{2}$$

Note When a device is biased half-way along the load line it is called *Class A biasing* (Fig. 3.3).

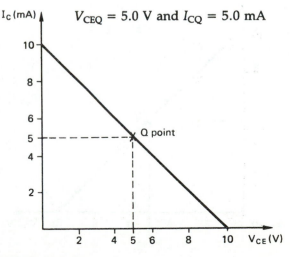

Fig. 3.3 Class A biasing

To consider this further let us look at the practical amplifier circuit shown in Fig. 3.4 and discover what exactly is going on.

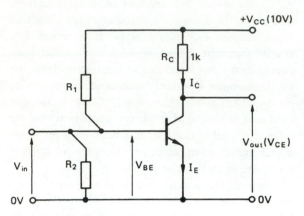

Fig. 3.4 The basic common emitter amplifier

The potential divider R_1, R_2 fixes the V_{BE} and hence the base current I_B. I_C is determined by I_B and the h_{FE} of the device. Remember that

$$I_C = h_{FE} \times I_B$$

V_{out} is V_{CE} and this is determined by the voltage drop across R_C caused by the current (I_C) flowing through it. So,

$$V_{CE} = V_{CC} - (I_C R_C)$$

Examine the load line for the circuit (Fig. 3.5).

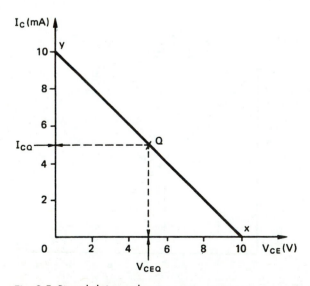

Fig. 3.5 Bias calculation graph

When $I_C = 0A$, $V_{CE} \simeq V_{CC}$
$\therefore V_{CE} = +10V$ (point x)

When $V_{CE} = 0V$,
$$\therefore I_C = \frac{V_{CC}}{R_C} = \frac{10V}{1k} = 10mA \text{ (point y)}$$

Now, where along this line is it sensible to bias the transistor for a Class A amplifier? The best place is approximately half-way along it. Here $V_{CEQ} = 5V$ (subscript Q indicates quiescence). If the transistor has an h_{FE} (min) of 250 then when

$$I_C = 5mA$$
$$I_B = \frac{I_C}{h_{FE}} = \frac{5 \times 10^{-3}}{250} = 20\mu A$$

So $I_{BQ} = 20\mu A$

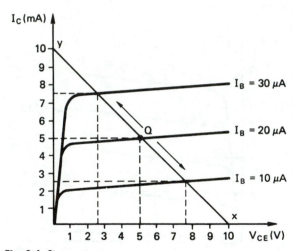

Fig. 3.6 Bias range

R_1 and R_2 are chosen to fix I_{BQ} at $20\mu A$, and so with no input signal

$$I_{BQ} = 20\mu A$$
$$I_{CQ} = 5mA$$
$$V_{CEQ} = 5V$$

because $V_{CE} = V_{CC} - (I_C R_C)$
$$= 10 - (5 \times 10^{-3} \times 1k)$$
$$= 10 - 5 = 5V$$

If an input signal is applied; as the input signal goes positive, I_B increases and point Q moves up the line towards y. Consequently I_C goes up, $\therefore (I_C R_C)$ increases and V_{CE} goes down.

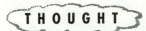

THOUGHT

- As the input voltage goes up the output goes down?
  Yes! Think about it: supposing the input signal causes I_B to rise to $30\mu A$. I_C will rise to $30\mu A \times 250 = 7.5mA$.

$$(I_C R_C) = 7.5 \times 10^{-3} \times 1k = 7.5V$$

so $V_{CE} = 10 - 7.5 = 2.5V$

Conversely as the input signal goes negative I_B will reduce and point Q moves towards x, so I_C goes down, $I_C R_C$ will decrease and V_{CE} will go up. The sequence is V_{in} positive, I_B and I_C up, V_{CE} down. V_{in} negative, I_B down, I_C down, V_{CE} up.

This is shown by the time related waveforms for an applied sine wave shown in Fig. 3.7.

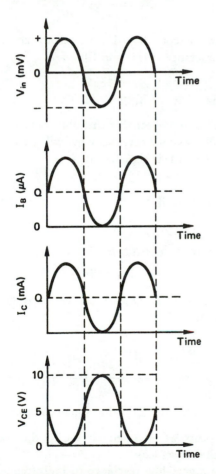

Fig. 3.7 Input/output characteristics of common emitter amplifier

The common emitter amplifier is thus an inverting amplifier; i.e. the output signal is the inverse (180° out of phase) of the input signal.

SELF ASSESSMENT 7

A Class A common emitter transistor amplifier has a 2k2 collector load resistance. If the collector current is 2mA under quiescent conditions and the supply voltage is 12V d.c. calculate:

1 The quiescent output voltage.
2 The maximum swing of the output voltage before distortion occurs.
3 The maximum swing of the output current before distortion occurs.
4 The quiescent base current if the h_{FE} is 120.

In essence the transistor can be used as a voltage amplifier. By selecting a suitable d.c. input current it can be turned partly on. Any a.c. input signal will vary this input current and the output current will also vary causing a varying output voltage.

There are a few requirements yet to be added to the circuit that ensure it will operate as a practical amplifier.

INPUT AND OUTPUT CAPACITORS (FIG. 3.8)

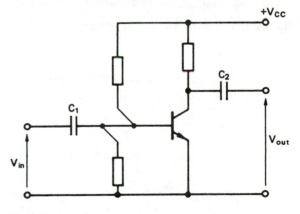

Fig. 3.8 Input and output capacitors

You are aware that the transistor is biased by fixing the operating point under quiescent conditions (no input signal). But supposing the circuit that supplies the signal into the amplifier has some d.c. present.

This 'signal' d.c. will shift the d.c. operating point of the transistor and upset the bias point (Fig. 3.9). The same kind of situation can exist with

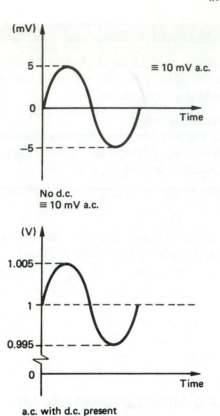

Fig. 3.9 Effect of additional d.c. on signal voltage

the output signal, e.g. for the circuit we have been considering the output looks like this:

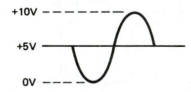

This is a varying d.c. signal, i.e. it swings about +5V. For a true a.c. output it should look like this:

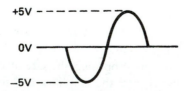

(swing about 0V). So for both input and output signals there is a need to separate the d.c. from the a.c. and allow only the a.c. through. A component

that does this most admirably is the humble capacitor.

SIGNAL COUPLING AND THE CAPACITOR

The capacitor is a reactive component. This means that the opposition it offers to an alternating signal is dependent upon two things:

1 the capacitance of the capacitor;
2 the frequency of the signal.

From studies you have done on electrical principles you will be aware that a capacitor's reactance is termed X_C and can be calculated using

$$X_C = \frac{1}{2\pi f C} \, \Omega$$

For example, consider a $10\mu F$ capacitor and calculate its reactance at 1Hz and 10kHz.

At 1Hz $X_C = 16k\Omega$
at 10kHz $X_C = 1.6\Omega$

It becomes very clear that the $10\mu F$ capacitor will pass a 10kHz signal very easily but will act as a high resistance to a 1Hz signal.

THOUGHT

■ *This is fine but what about d.c.?*
Well, consider the frequency of d.c.

$$d.c. = 0Hz$$

If you were to work through the reactance equation you would arrive at the result

$$X_C = \frac{1}{0} = \infty \, \Omega$$

A capacitor has the following signal characteristics.

1 It appears as an open circuit to d.c. signals (infinite resistance).
2 It has a very low reactance to high frequencies (low resistance).
3 It has a high reactance to low frequencies (high resistance).

It is important that the value of capacitor used for coupling input and output signals is chosen to have

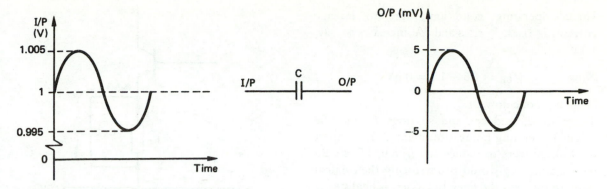

Fig. 3.10 Effect of a capacitor

a low reactance for the range of frequencies at which the amplifier operates.

A capacitor that allows the a.c. signals to pass while preventing the d.c. component from passing is referred to as either an *a.c. coupling capacitor* or a *d.c. blocking capacitor* (Fig. 3.10).

Thermal stability

It is important when considering transistors not to forget the minority carriers. These are always present, and under certain circumstances cause real problems. Recalling the diode, you will remember that the minority carriers gave rise to a leakage current under reverse bias conditions, and that this leakage current increased with temperature.

In a correctly biased transistor the base collector junction is reverse biased; this means that a leakage current will always be present, and that the actual collector current is made up of the wanted current flow *plus* the leakage current.

Now consider this:

$$I_E \simeq I_C$$

Under operating conditions I_C increases and so does the leakage. Therefore I_E increases. The device warms up slightly and the leakage current increases, so I_C increases and I_E goes up so the device gets hotter. Consequently the leakage increases, hence I_C increases and so does I_E, and so on. This cumulative action can bring about the destruction of the device if it remains unchecked.

So how can we control it? A heat sink will certainly prevent the device getting too hot but, and this is important, it will not stop temperature variations. Every time the temperature changes the leakage current will change and the operating point, which has been so carefully chosen, will move up the load line if the temperature increases, and down if it reduces.

The usual way to counteract this is to include a resistor R_E in the emitter circuit (Fig. 3.11).

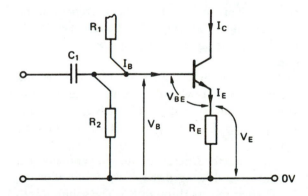

Fig. 3.11 Modified circuit showing resistor for thermal stability

R_1 and R_2 fix the base bias voltage V_B and hence the quiescent base current I_B. The V_{BE} is assumed to be 0.6V for a Si transistor.

If	$I_E \simeq I_C$
then	$V_E = (R_E I_C)$
If	$V_B = 1.6V$
and	$V_E = 1.0V$
	$V_{BE} = V_B - V_E = 0.6V$

Under operating conditions, due to leakage current, I_C rises, I_E rises and V_E increases to, say, 1.1V.

Now $V_{BE} = 1.6 - 1.1 = 0.5V$

(V_{BE} has gone down!)

Therefore I_B reduces and so does I_C and the original operating point is maintained. Likewise if the leakage were to reduce and I_C fall, V_E would reduce and V_{BE} would rise restoring the original state. In this way the operating point is held steady over a wide temperature range and the circuit is thermally stable.

This is a perfect example of negative feedback in action (see the section on negative feedback): any increase in the output current due to temperature affecting the leakage current is used to produce a feedback voltage (V_E) that reduces the input voltage V_{BE} and stabilises the circuit.

THERMAL STABILITY UNDER SIGNAL CONDITIONS

When the circuit is actually operating under signal conditions the base current is varying in accordance with the input signal and so is the collector current. This means that the current flowing through R_E is also varying and consequently V_E will be increasing and decreasing with the output current.

Now whenever V_E goes up, V_{BE} reduces and when V_E goes down, V_{BE} increases in order to keep the transistor thermally stable; if the a.c. signal varies V_E, the gain will be significantly reduced.

It is therefore necessary to prevent any a.c. current from flowing through R_E, while allowing d.c. current to pass through R_E. The solution is to use a capacitor to bypass R_E and thus take the a.c. signal to the 0V rail (Fig. 3.12). C_E is called a *bypass* or *decoupling capacitor* and is chosen to have a low reactance (low resistance) to the a.c. signals.

THOUGHT

■ *Won't some d.c. flow through C_E?*
No, because capacitors present an open circuit (infinite resistance) to d.c.

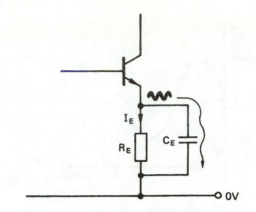

Fig. 3.12 The bypass (decoupling) capacitor

AN IMPORTANT POINT ABOUT DECOUPLING CAPACITORS

For the gain of the amplifier to be high, it is necessary to prevent any a.c. signals passing through R_E. Consequently the reactance of the emitter bypass capacitor is of much greater importance than that of the coupling capacitors. For example, a $10\mu F$ capacitor at a frequency of 1kHz has a reactance of 16Ω when used as a coupling capacitor. This is merely a series resistor of 16Ω and performs adequately, but if used as a decoupling capacitor (Fig. 3.13) it acts in parallel with the

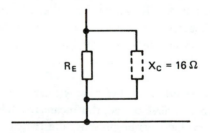

Fig. 3.13 Reactance of emitter bypass capacitor in circuit

resistor R_E and so whilst most of the signal will pass through C_E, some will still flow through R_E, significantly reducing the gain. The smaller the value of C_E, the lower the low frequency gain. For this reason C_E tends to be larger in practice than the coupling capacitors, typically $100\mu F$. For an audio amplifier

$$X_C \text{ at } 1kHz = 1.59\Omega$$

The complete practical transistor amplifier circuit is shown in Fig. 3.14.

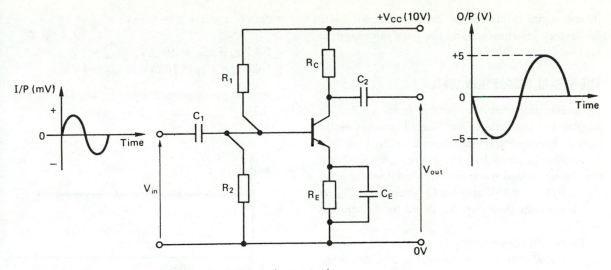

Fig. 3.14 Complete common emitter amplifier showing input and output signals

Amplifier characteristics

For a small signal amplifier the following are important.

- The Gain – the amount the input signal is increased by.
- The signal distortion level – the maximum input signal level before the output distorts.
- The bandwidth – the range of frequencies over which the amplifier operates.
- The input and output resistance – the resistance seen looking into the input and output terminals of the amplifier.

We shall examine these parameters individually in order to develop an understanding as to why they are so important.

THE GAIN

This refers to both A_i and A_v for a small signal amplifier. You will remember that when we considered BS notation for transistors the use of upper case subscripts indicated d.c. or static values (with no input signal connected). The use of lower case subscripts indicates that the value quoted is a small signal or a.c. value.

h_{FE} = d.c. forward current gain,
h_{fe} = a.c. or small signal current gain,

I_B = d.c. base current,
I_b = small signal base current.

It follows then that:

A_i = small signal current gain,
A_v = small signal voltage gain.

Since the d.c. values of current and voltage are established by the biasing used, it is understandable that any input signal applied to the amplifier will cause these d.c. values to change about their Q point. This will be a small change since we are considering small signals.

$$A_i = \frac{\text{small change in output current}}{\text{small change in input current}}$$

$$= \frac{\delta I_C}{\delta I_B} = \frac{I_c}{I_b}$$

where δ = the Greek letter delta ≡ small change,
I_C = output current as determined by the biasing,
I_B = input current, as determined by the biasing

likewise

$$A_v = \frac{\text{small change in output voltage}}{\text{small change in input voltage}}$$

$$= \frac{\delta V_{CE}}{\delta V_{BE}} = \frac{V_{ce}}{V_{be}}$$

These signal levels can be easily measured using laboratory instruments and the gain calculated, as will be described later.

THE SIGNAL DISTORTION LEVEL

It is essential that the output from a Class A amplifier is distortion free. The Q point is established approximately half-way along the load line. If the change caused by the input signal pushes it too far up or down the line the signal will distort, i.e. if $V_{CC} = +10V$ and the Q point fixes V_{CE} at $+5$ it can only swing up and down by a maximum of 5V.

There will consequently be a maximum value of input signal that can be applied to the amplifier. Anything greater than this will cause clipping of the output signal to occur (Fig. 3.15).

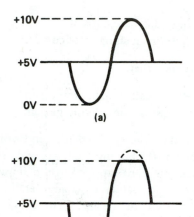

(a)

(b)

Fig. 3.15 Distortion due to 'clipping'

For example, an input signal of 40mV peak-to-peak may produce an undistorted output as in Fig. 3.15(a) whilst an input of 45mV may produce a distorted output as in Fig. 3.15(b).

∴ 40mV is the maximum input signal.

THE BANDWIDTH

An ideal amplifier would amplify all input signals by the same amount, regardless of their frequency

e.g. if $A_V = 100$ and $V_{in} = 10mV$, then at

$f = 0Hz$ (d.c.), if $V_{in} = 10mV$, $V_{out} = 1V$
$$(A_v \times V_{in})$$
$f = 10kHz$, if $V_{in} = 10$ mV, $V_{out} = 1V$
$f = 1MHz$, if $V_{in} = 10$ mV, $V_{out} = 1V$

A graph of gain against frequency would look like Fig. 3.16.

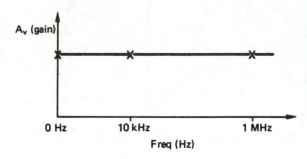

Fig. 3.16 Gain against frequency graph of ideal amplifer

Unfortunately the reality is somewhat different. Due to a number of factors, the gain which is quoted as 100 will change with the frequency of the input signal to such an extent that the amplifier will be quite useless above and below two critical frequencies. This is rather like the human ear which cannot hear frequencies below 20Hz and above 20kHz but is perfectly serviceable between these two extremes. The bandwidth (B) of any amplifier or system is the range of frequencies over which it can be used. It is specified as the upper and lower frequencies at which the gain has fallen to 70.7% of its mid-frequency gain value (Fig. 3.17).

Note This figure 70.7% is not due to any r.m.s. values.

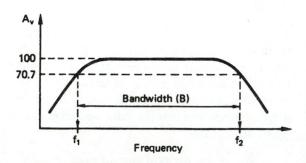

Fig. 3.17 Amplifier gain/frequency response curve

Mid-frequency gain = 100
∴ 70.7% of 100 = 70.7
f_1 = lower cut-off frequency
f_2 = upper cut-off frequency

For an audio amplifier f_1 and f_2 should be approximately 20Hz and 20kHz respectively. The amplifier will then have a similar response to the human ear.

THE INPUT AND OUTPUT RESISTANCE

An amplifier, however complex, has two input terminals and two output terminals. Any circuit, component or piece of equipment connected to the input or output of the amplifier 'sees' it as a simple resistor. For example, a signal generator connected to the input of an amplifier regards it as a resistor (R_{in}) in parallel with itself, as does a loudspeaker across the output (Fig. 3.18).

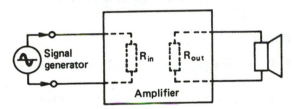

Fig. 3.18 Amplifier block diagram showing input and output resistance

This may appear to be unimportant until you realise that the result of connecting two resistors in parallel is a total resistance less than the smallest of the two.

Example Suppose the signal generator has an output resistance of 5k and the R_{in} of the amplifier is 500Ω.

$$R_T = \frac{R_{sig} \times R_{in}}{R_{sig} + R_{in}}$$

$$\therefore R_T = \frac{5000 \times 500}{5500} = 454\Omega$$

This is going to load the generator to such an extent that it may not work!

Suppose R_{out} of the amplifier is 120Ω and the loudspeaker has a coil resistance of 8Ω

$$R_T = \frac{120 \times 8}{120 + 8} = 7.5\Omega$$

This could draw excessive current through the amplifier and destroy it or at least cause severe signal distortion.

It is of paramount importance that the input and output resistances of an amplifier are known in order that the amplifier can be matched to any external circuit to ensure efficient operation. Methods of practically measuring these various parameters will be given in the next section.

Testing small signal amplifiers

It is important that all designers have a familiarity with the tools and processes that are to be used in the testing of any design. In the last section certain parameters or characteristics emerged as being significant;

■ the gain,
■ the signal distortion level,
■ the bandwidth,
■ the input and output resistance.

What follows next is an itemised guide to practically determining these factors using standard laboratory instruments. The next investigation uses a proven audio amplifier circuit so that you can develop your own skills by testing the amplifiers.

THE GAIN (Fig. 3.19)

This refers to the voltage gain A_v which you already know to be

$$A_v = \frac{V_{out}}{V_{in}}$$

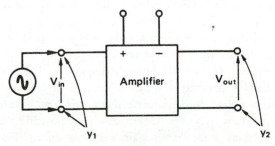

Fig. 3.19 Circuit to determine the gain

Method

1 Connect the amplifier to a power supply and adjust for the correct supply voltage.
2 Connect a signal generator to the input and adjust it to give a sine wave output of 1kHz at about 20mV peak-to-peak.

■ *Why 1kHz? This is a convenient frequency because it is not low enough to be influenced by the coupling capacitors, it is easy to get a clear stable trace on the CRO, and for audio amplifiers it is a standard test frequency.*

3 Monitor V_{in} and V_{out} with the CRO (check V_{out} is undistorted, if it is, reduce V_{in}).
4 Measure and record the peak-to-peak values from the CRO screen and calculate A_v.

THE SIGNAL DISTORTION LEVEL

1 Leave the arrangement as it was set up for the measurement of gain (Fig. 3.19).
2 While observing V_{out} on the screen, increase V_{in} (signal generator output) until V_{out} just starts to distort, i.e. clips or appears nonsinusoidal. Reduce V_{in} until V_{out} is just normal and measure V_{in}.
3 The value of V_{in} is now the maximum input signal that can be applied before distortion occurs.

THE BANDWIDTH

This is the range of frequencies over which an amplifier can be used. It is further specified as the range of frequencies over which the gain falls no lower than 70.7% of its mid-band value.

This can be a slightly tedious test to perform, but it is one of the most crucial, and if done correctly can be accomplished quickly. The test involves making a series of gain measurements from a very low frequency, say 10Hz, to a very high frequency, say 1MHz, and then plotting the results as a frequency gain response curve (Fig. 3.20).

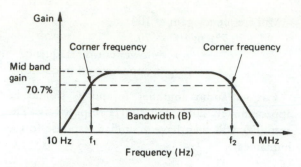

Fig. 3.20 Graph of gain against frequency

From this curve the upper and lower 'corner' frequencies f_1 and f_2 can be found and the bandwidth (B) established.

Using $B = f_2 - f_1$
if $f_1 = 25\text{Hz}$
and $f_2 = 980\text{kHz}$
then $B = 980\text{kHz} - 25\text{Hz} \simeq 980\text{kHz}$

Method

1 Set up the equipment as it was for measuring the gain (Fig. 3.19).
2 Set V_{in} to a value well below the maximum level, e.g. if V_{in} (max) = 50mV before distortion occurs, set it to 20mV (say).
3 Prepare a table thus:

Frequency	V_{in}	V_{out}	$A_v = \dfrac{V_{out}}{V_{in}}$
10Hz	20mV	24mV	1.2
30Hz	20mV	30mV	1.5

4 You are now going to make the series of measurements that will enable the graph to be drawn.

■ *Over what frequency steps are the measurements made? If the amplifier has a flat response then it is only necessary to make a large number of measurements when the gain is changing (Fig. 3.21).*

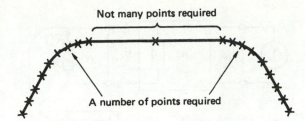

Fig. 3.21 Illustration of the method of plotting a gain-frequency graph

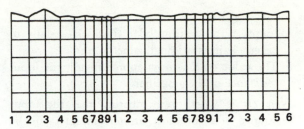

Fig. 3.22 Three cycle log/linear graph paper

So it is a good idea to sweep or scan the entire range from 10Hz–1MHz with the generator before recording any results just to see how the gain changes. Then from this broad overview you can decide where over the range to make the most measurements.

5 Make your measurements, calculate A_v and plot the curve.

Note It is, I think, a good idea to plot the graph as you go. In this way any 'rogue' measurements will be immediately apparent and, since the equipment is connected, checking and repeating is simple. Otherwise you will be in the position of plotting your results from a table, and any errors will mean considerable work if they are to be corrected!

Plotting the curve

You will need log/linear graph paper for this, with the logarithmic scale along the horizontal axis and the linear scale vertical. This type of paper is available in cycles, e.g. 3 cycle log/linear paper (Fig. 3.22).

The reason for this cramping of the frequency scale is to enable a large frequency range to be accommodated on a reasonably sized sheet of paper. Notice the lowest figure is 1 and the highest 9. This means you decide on the actual values, e.g. 1 could be 1Hz then 9 would be 9Hz the next 1 would be 10Hz 2 would be 20Hz etc.

So if you are doing a response over the range 10Hz–1.5mHz you will need 6 cycle log/lin paper:

10Hz 90Hz	100Hz 900Hz	1kHz 9kHz	10kHz 90kHz	100kHz 900kHz	1MHz 9MHz
1 cycle	2 cycles	3 cycles	4 cycles	5 cycles	6 cycles

The gain once calculated is plotted up the side normally. When your points have been transferred

they can be joined to form the curve, the mid-band gain can then be established, let's say 48.

The bandwidth (*B*) is enclosed by the upper and lower frequency points where the gain is 70.7% of 48 (48 = mid-band gain).

f_1 and f_2 = 0.707 × 48 ≃ 34 (70.7% of mid-band gain)

B can now be firmly established.

INPUT AND OUTPUT RESISTANCE

Perhaps the easiest method of estimating these is to use Ohm's law and a variable resistance, preferably a decade resistance box in conjunction with the usual laboratory signal generator and oscilloscope. You will do well to remember that the accuracy of the result reflects the care and vigilance with which the testing has been carried out.

Input resistance (Fig. 3.23)

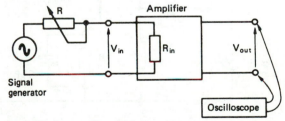

Fig. 3.23 Circuit to determine input resistance

1 Set R to zero (0Ω = short circuit).
2 Set the signal generator to 1kHz sine wave and adjust to give a V_{in}, i.e. well below the maximum permissible level before distortion.

3 Measure V_{out}.

4 Increase the setting of R until V_{out} falls to exactly half its original value.

5 The value of R is now approximately the input resistance of the amplifier.

Note It is worth remembering that the input resistance of a common emitter amplifier will be approximately the quoted h_{ie} of the transistor. For a common source amplifier R_{in} will be approximately that of the gate resistor R_G.

Output resistance (Fig. 3.24)

1 Disconnect R and monitor V_{out} with the CRO.

2 Connect the signal generator and adjust to give a suitable value of V_{out} at 1kHz.

3 Connect R and adjust until V_{out} is exactly half the value it was.

4 The value of R now represents the output resistance of the amplifier.

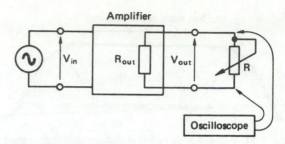

Fig. 3.24 Circuit to determine output resistance

Note This is a particularly difficult test to perform because of the comparatively low resistance of the amplifier output. If any distortion occurs in the output signal repeat the test with a lower value of V_{in}.

Now, with this information available, perform Practical Investigation 15 and produce your own set of test results for an audio amplifier.

PRACTICAL INVESTIGATION 15

Testing a small signal audio amplifier

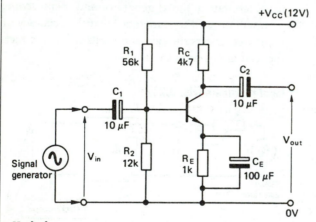

Equipment

n-p-n general purpose transistor e.g. BC108
56k, 12k, 4k7, 1k resistors
Two 10μF capacitors
100μF capacitor
Dual beam oscilloscope (CRO)
Power supply
Signal generator
Decade resistance
6 cycle log/lin graph paper
Digital multimeter

Method

1 Build the circuit shown above.

2 Connect the power supply and signal generator. Monitor V_{in} and V_{out} with the CRO and check that the circuit amplifies.

3 With no input signal, measure and record the d.c. bias voltages on the base, collector and emitter with respect to the 0V line.

4 Connect the signal generator and adjust to give a 1kHz sine wave.

5 Using the procedures discussed previously carry out the following tests:

   a) gain,   **b)** signal distortion level,   **c)** bandwidth,
   d) input resistance,   **e)** output resistance.

In each case be methodical and record your results in a clear and understandable manner.

Results

At the end of your deliberations you should be in possession of detailed specifications for the amplifier under test, including a visual indication of its frequency response performance in the form of a graph.

Feedback

It is almost impossible to consider the topic of amplifiers without the term 'feedback' being mentioned. Indeed the concept has previously been broached when the subject of thermal stability was considered.

Where amplifiers are concerned feedback is the art of taking a portion of the output signal and *feeding-it-back* to the input, this is illustrated in Fig. 3.25.

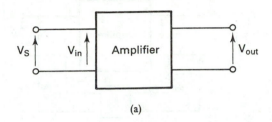

(a)

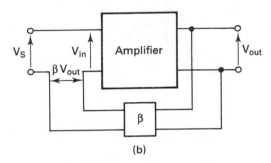

(b)

Fig. 3.25 (a) Amplifier without feedback (open loop). (b) Amplifier with feedback applied

Fig. 3.25a shows an amplifier without any feedback (sometimes called an 'open-loop' amplifier). Here the input signal (V_s) is the actual input to the amplifier (V_{in}).

$$\therefore V_s = V_{in}$$

Fig. 3.25b shows the same amplifier with feedback applied. You will see that there is a feedback network (this may be a simple resistor!) that takes a fraction of the output signal and applies it to the input of the amplifier. This means that the actual input to the amplifier is now the input signal plus the feedback signal. This can be expressed in the following way:

V_s = input to the system
V_{in} = input to the amplifier
V_{out} = output of the amplifier
β = feedback fraction

The signal that is fedback to the input will be:

$$\beta V_{out}$$

So the input to the amplifier itself (V_{in}) will now be:

$$V_{in} = V_s + \beta V_{out}$$

Negative and positive feedback

We have identified the feedback signal as βV_{out}. If this is a positive voltage ($+\beta V_{out}$) then:

$$V_{in} = V_s + (+\beta V_{out})$$
$$\therefore V_{in} = V_s + \beta V_{out}.$$

The input signal V_{in} will increase, this is the result of '*positive feedback*'. If however βV_{out} is negative ($-\beta V_{out}$) then:

$$V_{in} = V_s + (-\beta V_{out})$$
$$\therefore V_{in} = V_s - \beta V_{out}$$

Thus the input signal V_{in} will now reduce, this is '*negative feedback*'.

This was precisely the situation when thermal stability was considered in the previous section, as the temperature of the device increased the output increased and a *negative feedback* voltage was produced that was subtracted from the input voltage, thus making it smaller. This in turn reduced the output and the circuit became more stable.

Signal (a.c.) feedback

Amplifiers may be used to amplify alternating signals, e.g. audio frequencies. It is usual to use a.c. or signal feedback to modify the performance of the amplifier. We now have to consider positive and negative feedback in respect of these signals. Consider the case of positive a.c. feedback as depicted in Fig. 3.26a and the associated waveforms shown in Fig. 3.26b.

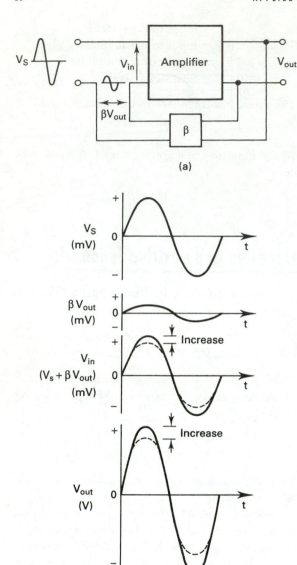

Fig. 3.26 (a) Amplifier with positive feedback applied. (b) Effects of positive feedback

Here you will see that the feedback signal is '*in phase*' with the input signal V_s and so V_{in} increases, this has the effect of increasing V_{out} and hence the size of the feedback signal.

THOUGHT

■ If an increase in V_{out} causes an increase in βV_{out} then V_{in} goes up and so does V_{out}, which in turn increases βV_{out} etc, is this right?

Yes! This will result in instability and is the reason why positive feedback is a thing to be avoided in amplifiers.

Now consider the case of negative a.c. feedback as shown by Fig. 3.27a and the wave forms in Fig. 3.27b.

The feedback signal here is in '*anti-phase*' to the input signal (V_s). This means there is a 180° phase

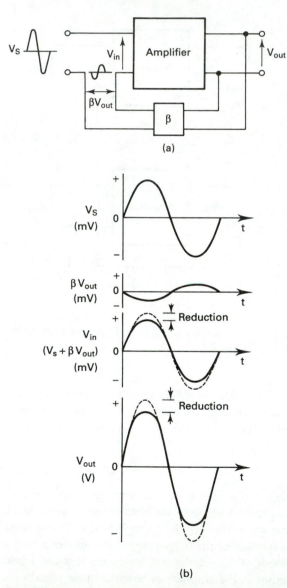

Fig. 3.27 (a) Amplifier with negative feedback applied. (b) Effects of negative feedback

shift between input and output. This negative feedback signal will cause V_{in} to reduce and hence V_{out} will reduce as shown in Fig. 3.27b.

THOUGHT

■ *If V_{out} decreases then the feedback will decrease and so will V_{in}. Surely the output will eventually fall to zero since this is the opposite to positive feedback?*
No! Think carefully about this, βV_{out} causes V_{in} to decrease slightly, this in turn reduces V_{out}, this will reduce βV_{out} which will allow V_{in} to rise slightly. Far from continually reducing the output the application of negative feedback will keep the gain reasonably constant, i.e. it will stabilise the amplifier.

Quick recap

Positive feedback produces instability in an amplifier and is consequently avoided.
Negative feedback improves an amplifier's stability and modifies the performance so it is frequently used.

The effects of negative feedback

GAIN

We have seen that negative feedback reduces the output signal (V_{out}) but since the input signal to the system (V_s) is the same, the overall gain of the amplifier will have been reduced. We can now make the statement that the application of negative feedback *reduces* the gain of an amplifier.

BANDWIDTH

If the gain has been reduced the bandwidth must also change – see Fig. 3.28.

You may see from this that if the gain is reduced the bandwidth will increase, therefore negative feedback reduces the gain but widens the bandwidth of an amplifier. This phenomena is illustrated by Practical Investigation 16.

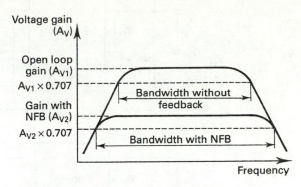

Fig. 3.28 Effect of NFB on gain and bandwidth

Input and output impedance

The fact that a feedback circuit or component is connected to both the output and the input of an amplifier means that the input and output impedances will change as a result. There is more than one way of applying feedback to an amplifier and Fig. 3.29 illustrates this.

So the feedback signal can be either a current or a voltage and it can be applied in series or parallel (shunt) to the input.

Each of these modes of connection will affect the impedances in different ways. It is sufficient at this stage simply to accept that negative feedback changes the input and output impedances of an amplifier.

GAIN STABILITY

As previously outlined the application of negative feedback improves the stability of an amplifier.

NOISE AND DISTORTION

Since the gain is reduced by negative feedback the amount of noise and distortion produced by the amplifier is also reduced.

a.c. AND d.c. FEEDBACK

You will remember that d.c. negative feedback is used to provide thermal stability and the use of a bypass or decoupling capacitor shown in Fig. 3.30a prevents a.c. negative feedback occurring. It is possible to provide a.c. or d.c. negative feedback

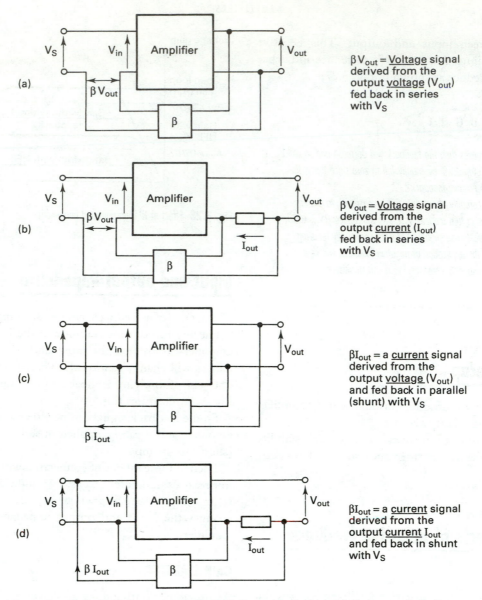

βV_{out} = <u>Voltage</u> signal derived from the output <u>voltage</u> (V_{out}) fed back in series with V_S

βV_{out} = <u>Voltage</u> signal derived from the output <u>current</u> (I_{out}) fed back in series with V_S

βI_{out} = a <u>current</u> signal derived from the output <u>voltage</u> (V_{out}) and fed back in parallel (shunt) with V_S

βI_{out} = a <u>current</u> signal derived from the output <u>current</u> I_{out} and fed back in shunt with V_S

Fig. 3.29 (a) Series voltage feedback. (b) Series current feedback. (c) Shunt voltage feedback. (d) Shunt current feedback

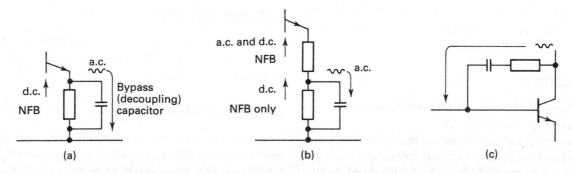

Fig. 3.30 (a) d.c. negative feedback only. (b) Partial a.c. negative feedback. (c) a.c. negative feedback only

PRACTICAL INVESTIGATION 16

Negative feedback

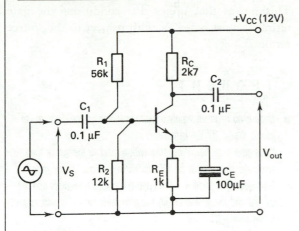

Equipment
n-p-n general purpose transistor
56k, 12k, 2k7, 1k resistors
Two 0.1 µF capacitors
100 µF capacitor
Dual beam oscilloscope (CRO)
Power supply
Signal generator
6 cycle log/lin graph paper

Method
1 Build the circuit shown above.
2 Connect the power supply and signal generator. Monitor V_s and V_{out} with the CRO and check that the circuit amplifies.
3 Using the procedures outlined previously carry out the following tests and provide the following results.
   a) Voltage gain (A_v).
   b) Frequency response (plotted as A_v against frequency).
   c) From your frequency response graph determine the bandwidth (B).
4 Disconnect capacitor C_E so that a.c. negative feedback occurs and repeat the tests plotting the frequency response curve on the same axis.

Results
Compare the results from both tests and verify that the statements previously made about the effects of negative feedback on voltage gain and bandwidth are true in practice.

---

or both as the occasion demands. Figure 3.30b shows a.c. and d.c. negative feedback while Fig. 3.30c shows the application of a.c. feedback only, the capacitor effectively blocking any d.c. component.

THOUGHT

■ *What happens if the gain of the amplifier is too low as a result of negative feedback?*
The answer is to use more than one amplifier stage. In this way it is possible to design a two- or three-stage amplifier and, by varying the amount of negative feedback applied to each stage, customise the circuit to produce exactly the gain and bandwidth required.

The unipolar transistor amplifier

THE JUNCTION FET

From previous work on the JFET you know that it operates in depletion mode; the channel can be narrowed by the bias voltage on the gate. To use it as an amplifier it is required that the device is biased somewhere between 'off' and 'on' under quiescent conditions. So considering an n-channel device the load line might look something like Fig. 3.31.

For a Class A amplifier a suitable Q point would be that provided by $V_{GS} = -3.0V$, giving $V_{DSQ} = +5V$ and $I_{DQ} = 10mA$. Any input signal connected to the amplifier would cause the

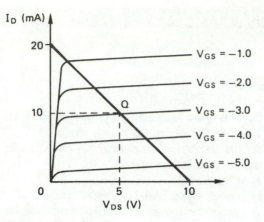

Fig. 3.31 Load line of a JFET

Q point to move just as it did for the BJT. So under signal conditions the voltage gain

$$A_v = \frac{V_{ds}}{V_{gs}} = \frac{\delta V_{DS}}{\delta V_{GS}}$$

$$= \frac{\text{small change in output voltage}}{\text{small change in input voltage}}$$

and the current gain

$$A_i = \frac{I_d}{I_g} = \frac{\delta I_D}{\delta I_G}$$

$$= \frac{\text{small change in output current}}{\text{small change in input current}}$$

But you will remember that the gate current for an FET is incredibly small. Consequently the amplifier current gain is not usually considered or given a finite value.

The circuit diagram for an n channel FET amplifier is shown in Fig. 3.32.

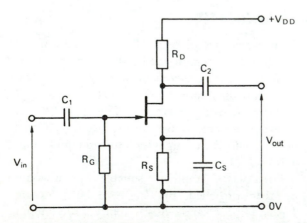

Fig. 3.32 The n channel JFET common source amplifier

Biasing

To establish the Q point the load line is used and the appropriate value of V_{GS} chosen (say $-3V$). This means that under d.c. conditions the gate terminal must be $-3V$ with respect to the source terminal.

THOUGHT

■ There is no negative supply or voltage shown on the diagram, so how is the $-3V\ V_{GS}$ achieved?
If the gate must be 3V negative with respect to the source then surely the source must be 3V positive with respect to the gate? Consequently if there is +3V between the source terminal and the 0V line and the gate is at 0V, the gate must be $-3V$ with respect to the source (Fig. 3.33).
So to fix the Q point it is only necessary to positively bias the source by the amount of V_{GS} required. Therefore R_S is the bias resistor.

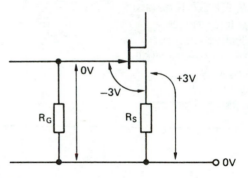

Fig. 3.33 Biasing the gate

THOUGHT

■ What is the purpose of R_G?
It is important that the gate appears to be at 0V, otherwise the bias will not be correct, i.e. if the source is at +3V with respect to the 0V line, the gate must appear to be at 0V or V_{GS} will not be $-3V$.
Obviously the gate cannot be linked to the 0V line by a piece of wire or V_{GS} would not be able to vary with any input signal applied. However the gate current for a FET is incredibly small. Therefore any resistor connecting the gate to the 0V rail will have only this tiny current flowing down it and will consequently drop a minute voltage across it, with the result that if one end of R_G is connected to the 0V rail, the gate will also be at 0V. So, if the gate is at 0V and the source at +3V the gate is $-3V$ with respect to the source.

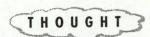

THOUGHT

■ *Why is R_G so large, e.g. 1MΩ?*
One of the big advantages of the FET is its high input impedance. If R_G were a low value, it would counteract this benefit. Consequently it is usually as high as possible.

Capacitors C_1 and C_2

As with the BJT amplifier, these are coupling capacitors that allow only a.c. signals to pass to and from the amplifier. Any d.c. is effectively blocked. Consequently they are chosen to have a low reactance at the required signal frequencies.

Capacitor C_s

This is a decoupling capacitor ensuring that no a.c. signals pass through the bias resistor R_S. If they did, the biasing of the FET would change with the a.c. signal; this would mean a.c. negative feedback resulting in much reduced gain (A_v).

It is worth noting that the FET is more thermally stable than the BJT due to its having fewer minority carriers present in the channel. This means it is not necessary to take design steps to prevent thermal runaway, although of course heat sinks must still be used if the device is operated near its specified limits.

THE BJT TRANSISTOR MODEL

In an amplifier the output current (I_c) is related to the input current (I_b) (lower case subscripts denote small signals), but the reality is that it is the input voltage signal that changes (I_b) and hence the output current. It makes sense to relate the output current to the input voltage and this can be done using a simple 'small signal' model of the transistor.

There are two electrical principles theorems that assist us in this matter: Thevenin's theorem and Norton's theorem. These are dealt with in detail in books about electrical principles, but I have outlined the basics below.

THE THEVENIN EQUIVALENT CIRCUIT

Any series, parallel network of resistors, however complicated, can be represented by a voltage source of internal resistance R_o, feeding a single equivalent resistor (R_{eq}). Fig. 3.34a shows a resistor network and Fig. 3.34b shows the Thevenin equivalent circuit.

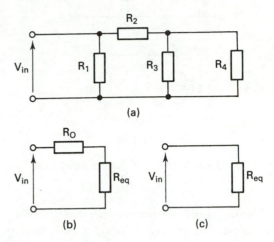

Fig. 3.34 (a) Series, parallel network. (b) Thevenin equivalent circuit. (c) Simplified equivalent circuit

Note If R_o is very small compared with R_{eq} the circuit further simplifies to that shown in Fig. 3.29c.

The input resistance of a transistor is often specified by the manufacturers in their detailed data sheets. This is quoted as the device's h_{ie} (input resistance in common emitter mode).

This means that the input section of the transistor can be represented as a voltage source V_{in} feeding a single resistor R_{in} as shown in Fig. 3.35.

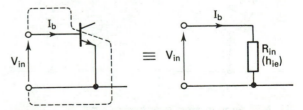

Fig. 3.35 Thevenin equivalent circuit of the input to a transistor

THE NORTON EQUIVALENT CIRCUIT

Here a network can be represented by a source that is a *constant current generator* feeding an equivalent resistor (R_{eq}). Fig. 3.36a shows a resistor network and Fig. 3.36b is the Norton equivalent circuit.

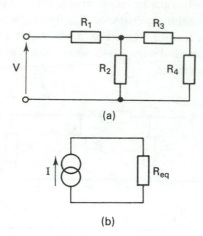

(a)

(b)

Fig. 3.36 (a) Resistor network. (b) Norton equivalent circuit

The transistor is a current operated device with the output current (I_c) a function of h_{fe} and the input current I_b.

$$h_{fe} = \frac{I_c}{I_b}$$
$$\therefore I_c = h_{fe} \times I_b$$

It is thus convenient to represent the output of the transistor using the Norton equivalent circuit of Fig. 3.37, i.e. a constant current generator feeding the output resistance (R_o) of the transistor.

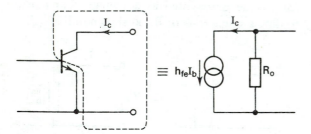

Fig. 3.37 Norton equivalent of the transistor output

This means that a transistor can be modelled or represented by its Thevenin and Norton equivalent circuits, i.e. by showing it is an input resistance R_{in} and current generator as shown in Fig. 3.38.

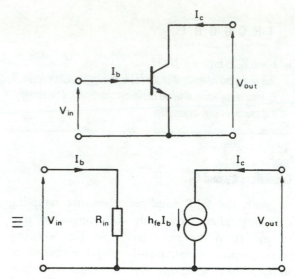

Fig. 3.38 Representation of a transistor as an input resistance and a current generator

Note R_o of a transistor is usually very high so it can be left out of the simplified equivalent circuit.

Determining values

The input current I_b is caused by V_{in} (V_{be}) and we now wish to relate I_c to V_{be}. This can be achieved using a term called the *forward transconductance figure* (*gm*) for the device (sometimes called the *mutual conductance*) which relates output current to input voltage

$$gm = \frac{I_c}{V_{be}}$$

(the units are A V^{-1}, called Siemen (S)).
   The *gm* for most transistors has a value of 40mS for every milliampere of collector current I_c. Therefore,

if $gm = 40\text{mS}$
                        $I_c = 1\text{mA}$

For practical purposes this can be expressed as

$$gm = \frac{I_c}{25} \text{ (mS)} \qquad \text{(where } I_c \text{ is in mA)}$$

For example, if $I_c = 1\text{mA}$, $gm = \dfrac{1}{25} = 40\text{mS}$

$$I_c = 8\text{mA}, \ gm = \frac{8}{25} = 320\text{mS}$$

This is most impressive, but what's the point?

The point is that this transistor model can be extended to represent a complete amplifier circuit and then used to estimate the voltage gain that the practical amplifier will have. Consider the practical amplifier shown in Fig. 3.39.

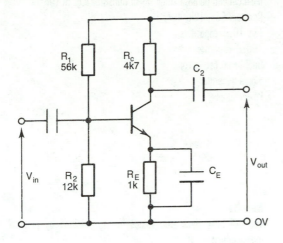

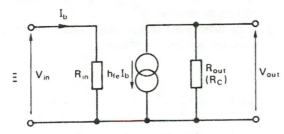

Fig. 3.39 Equivalent circuit of a practical amplifier

R_1 and R_2 together with the transistor input resistance h_{ie} make up the total input resistance R_{in}. The output resistance can be considered to be the collector load resistance R_C. The result of this is that the voltage gain A_v of the amplifier can be approximated to:

$$A_v = -gm \times R_C$$

Note The minus sign simply indicates that the amplifier is an inverting amplifier (180° phase shift) between input and output signals.

Example 1

A transistor used as an amplifier in common emitter mode has a collector load resistor of 3.9k. Under signal conditions $\delta I_C = 5$mA. Calculate the voltage gain A_v.

$$gm = \frac{I_c}{25} \text{ (small signal current)}$$

$$\delta I_C = I_c$$

$$\therefore gm = \frac{5}{25} = 200\text{mS}$$

$$R_c = 3k9$$

$$A_v = -gm \times R_C$$

$$\therefore A_v = 200 \times 10^{-3} \times 3.9 \times 10^3$$

$$= -780$$

Example 2

An amplifier has a 5Vpk−pk output signal V_{out}. If the collector load resistor R_c is 1k, calculate gm and hence the voltage gain.

$$\text{If} \quad V_{out} = 5\text{Vpk} - \text{pk}$$

$$\delta V_{CE} = 5\text{Vpk} - \text{pk}$$

$$\therefore V_{ce} = 5\text{Vpk} - \text{pk}$$

$$I_c = \frac{V_{ce}}{R_c} = \frac{5\text{V}}{1000} = 5\text{mA}$$

$$\therefore I_c = 5\text{mA}$$

$$gm = \frac{I_c}{25} = \frac{5}{25} = 200\text{mS}$$

$$A_v = -gm \times R_c = -200\text{mS} \times 1000 = -200$$

The FET model

When studying the unipolar device you became aware that one of the main parameters is gm (Y_{fs} or g_{fs}). This means the factor gm that had to be calculated for the BJT is quoted in the manufacturer's specification for an FET. The reason for this is that the FET is a voltage operated device in which the output current is controlled by the input voltage. For a BJT we had to convert the input current into input voltage, because it is a current operated device.

The model for an FET under signal conditions is similar to a BJT (Fig. 3.40), where R_{ds} is the actual resistance between the drain and source electrode of the device. The current generator ($gm V_{gs}$) produces a current I_d that is determined by the input voltage

$$I_d = gm \times V_{gs}$$

and

$$V_{in} = V_{gs}$$

PRACTICAL INVESTIGATION 17

Verification of the BJT transistor model

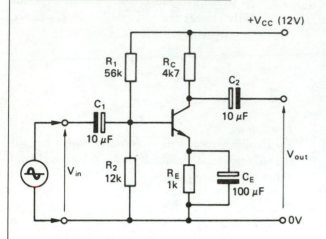

Equipment
n-p-n general purpose small signal transistor e.g. BC108
Resistors 56k, 12k, 4k7, 1k
Two 10μF capacitors
100μF capacitor
Dual beam oscilloscope
Signal generator
Power supply

Method
1 Build the circuit shown above and connect the power supply.
2 Monitor V_{in} and V_{out} with the CRO.
3 Connect the signal generator and adjust it to give a 1kHz sine wave input signal that produces an undistorted output.
4 Calculate the voltage gain A_v using $A_v = \dfrac{V_{out}}{V_{in}}$.
5 Repeat this with two other values of V_{in} at 1kHz ensuring that the output is undistorted.

Results
You will have three true values for A_v obtained by practical measurement.
6 Calculate the value of A_v using $A_v = -gmR_c$ and compare these with the real value.
7 Do the theoretical and practical values compare? Think of reasons that may explain any differences.

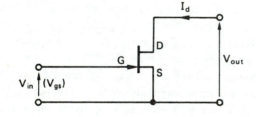

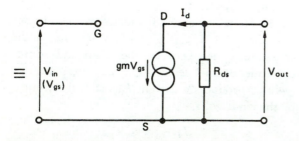

Fig. 3.40 FET small signal model

> ## THOUGHT
>
> ■ Why is R_{in} not shown?
> This is because the input resistance of an FET is very high and so under signal conditions can be ignored. If this model is extended to incorporate the complete amplifier then it looks like Fig. 3.41. R_{in} consists of R_G whilst R_{out} is made up of the transistor's total output resistance R_{ds} and the drain load resistor R_D.

As with the BJT amplifier the voltage gain of the circuit can be estimated using

$$A_v = -gm \times R_D$$

(remember the minus indicates the amp is inverting).

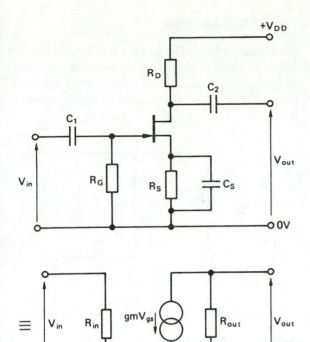

Fig. 3.41 Equivalent circuit of the common source amplifier

Example A BF245A junction FET is to be used as a Class A amplifier. Estimate the voltage gain A_v if the drain load resistor is

a) 4k7, b) 12k, c) 18k.

a) $A_v = -gm\,R_D$
$-gm = 3$ mS (min) from the data sheet
$\therefore A_v = -3 \times 10^{-3} \times 4.7 \times 10^{3}$
$= -14.1$

b) $A_v = -3 \times 10^{-3} \times 12\text{k}$
$= -36$

c) $A_v = -3 \times 10^{-3} \times 18 \times 10^{3}$
$= -54$

A word of warning

As with many methods of arriving at theoretical answers to practical problems, the estimated voltage gain may be quite different from the actual gain that is the result of building and testing the amplifier. This is because the calculations we have been using are only *approximate*. Also, the result is obtained by using theoretical values. In practice, the resistors used will have actual values according

to their tolerances, and the transistors themselves will have values slightly different to those quoted in the data. Finally, when the circuit is built and tested the signal generator connected to the input will slightly change the amplifier input resistance just as the oscilloscope will affect its output resistance.

So, the whole point of theoretical calculations and using models is to establish a starting point. Electronics is no different from any other discipline in this respect. The aeronautical designer calculates and arrives at a theoretical shape; he or she must then build a test piece to establish the actual practicalities of the theoretical design.

Other types of practical amplifiers

THE TUNED AMPLIFIER

From the section on transistor modelling it is apparent that the voltage gain of an amplifier is governed by the collector or drain load resistor

$$(A_v = -gm\,R_C, A_v = -gm\,R_D)$$

Now, if this resistor were replaced by a parallel tuned circuit (Fig. 3.42), the load resistance would be frequency dependent. The tuned circuit comprising C and L will have a resonant frequency (f_o) given by

$$f_o = \frac{1}{2\pi\sqrt{LC}}\ \text{Hz}$$

At this frequency the impedance (resistance) will be maximum, e.g. if C = 0.1μF and L = 1mH

$$f_o = \frac{1}{6.28\sqrt{0.1 \times 10^{-6} \times 1 \times 10^{-3}}} = 15.923\text{KHz}$$

So, at this frequency the load resistance will be maximum and A_v will be at its largest. At frequencies above and below f_o the resistance will be lower and so will the gain (because the gain of an amplifier is proportional to the load resistance). This will give rise to a gain frequency response curve that looks like Fig. 3.43.

Due to the response of the tuned circuit this gives a very narrow bandwidth amplifier that can be described as *selective*. This is particularly useful

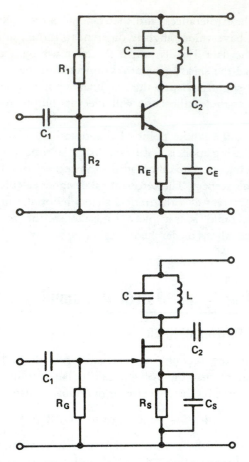

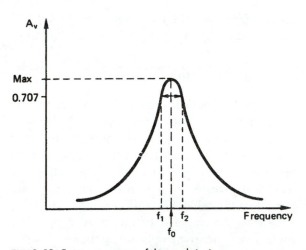

Fig. 3.42 Tuned amplifier with a frequency dependent load

for amplifying a chosen frequency or frequencies whilst neglecting all others. One common application is in the tuning stages of radio receivers.

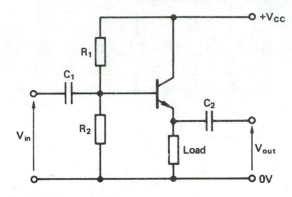

Fig. 3.43 Frequency response of the tuned circuit

THE CURRENT AMPLIFIER

If a transistor is used in emitter follower or source follower mode (also known as common collector and common drain) it has a very low voltage gain ($A_v \simeq 1$) but it has current amplification properties (Fig. 3.44).

Fig. 3.44 Current amplifier

The device can be used to act as an intermediate stage when it is required to drive a load that needs more current than can be supplied by an external circuit. The benefits of this mode of connection are:

1 high input, low output resistance,
2 it does not invert the signal,
3 it provides current amplification.

Typical application (Fig. 3.45)

TR_1 cannot drive the loudspeaker itself so TR_2 acts as an intermediate stage, 'buffering' TR_1 from the load. Amplifiers of this type are often called *buffer* amplifiers.

Designing small signal amplifiers

This is perhaps the 'acid test'. Having experimented and investigated the way transistors and amplifiers behave you now have to use the skills and knowledge acquired to design a circuit to perform certain specified functions. Designing electronic circuits is similar in many ways to designing anything; a set of specifications is produced based on what the finished product should be able to do, and a theoretical design is created. A prototype is made

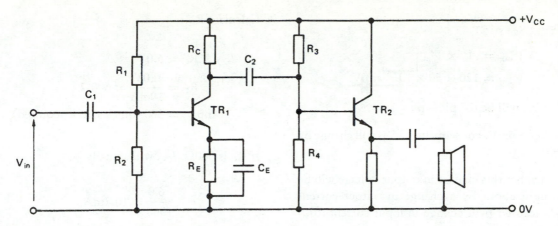

Fig. 3.45 Typical application of a current amplifier

DESIGN ASSIGNMENT 4

Amplifier (1)

A common emitter Class A audio amplifier is to be designed that must meet the following specifications:

1 Power supply – 12V d.c.
2 Voltage gain – 120 when V_{in} is a 1kHz sine wave of 50mV peak-to-peak.

3 Bandwidth – 500Hz–20kHz (minimum).

In addition to the above specifications a full test report will be required that includes all information relating to such things as

a) d.c. voltage levels
b) signal distortion level
c) input and output resistance

---

and tests are run to see if it comes up to the specifications. If it does not, modifications and improvements are made consisting of minor changes in the original design and the tests repeated. This process is continued until a satisfactory situation exists.

When designing an amplifier you will have at your disposal the results of your previous practical investigations. These, together with the modelling approach and the section that follows, will mean that you have the tools to do the job, all that remains is to flex your muscles and produce the goods. . . .

Procedure

Starting point
Sketch the proposed circuit then start with that which is already known and develop the unknown.

1 Determine V_{out}.

$$V_{in} = 50\text{mV pk} - \text{pk at 1kHz}$$
$$A_v = 120$$

Since $\quad A_v = \dfrac{V_{out}}{V_{in}}$

$$V_{out} = A_v \times V_{in}$$

$$V_{out} = 120 \times 50 \times 10^{-3} = 6V$$

$\therefore V_{out}$ will be 6V pk − pk

$\delta V_{CE} = 6.0$V, consequently V_{CE} will change by ± 3V.

Remember this change in output voltage will be brought about by a change in output current (δI_C) caused by a change in input current (δI_B) produced by V_{in}.

2 Decide on a suitable quiescent output voltage V_C.

Since $+V_{CC} = 12$V, $V_C \simeq \dfrac{V_{CC}}{2} = 6$V

Now $V_C = V_{CEQ} + V_E$

3 Decide on a value for V_E.
A rule of thumb is that V_E should be low compared with V_{CEQ} – so let's make $V_E = 1$V.

4 Now we can start calculating the biasing of the transistor in order to fix the Q point giving I_{CQ}.

Since $V_E = 1$V and $V_C = 6$V

$$V_{CEQ} = V_C - V_E = 6 - 1 = 5V$$

Now

$$V_{CEQ} + V_E = V_{CC} - I_{CQ} R_C$$
$$\therefore I_{CQ}R_C = V_{CC} - (V_{CEQ} + V_E)$$
So $\quad I_{CQ}R_C = 12 - (5 + 1) = 6V$

$I_{CQ}R_C$ should be about 6V

5 Determine I_{CQ}.
The type of transistor used will help with this decision, and as this is a small signal amplifier almost any general purpose transistor will be appropriate. Let's say that a BC108 is readily available. Refer to data.

$$I_C \text{ (max)} = 100\text{mA}$$

So make I_{CQ} a reasonable value that's easy to work with, e.g. 10mA:

$$I_{CQ} = 10\text{mA}$$

6 Calculate R_C.

$$I_{CQ} \times R_C = 6.0V$$

$$R_C = \frac{6.0V}{10mA} = 600\Omega$$

n.p.v. = 560 (quick check required)

If R_C is 560Ω, I_{CQ} will actually be

$$\frac{6.0}{560} = 10.7\text{mA}$$

This is OK, so $R_C = 560\Omega$

7 Calculate R_E.

$$V_E = 1.0V$$

So $\quad R_E = \dfrac{V_E}{I_{EQ}}$

Since $\quad I_{EQ} \simeq I_{CQ}$

$$R_E = \frac{V_E}{I_{CQ}} = \frac{1.0}{10.7\text{mA}} = 93.5$$

n.p.v. = 100Ω

This will give an actual

$$V_E = 100 \times 10.7\text{mA} = 1.07V$$

So $R_E = 100\ \Omega$

8 Calculate the quiescent base current I_{BQ}.

$$I_{BQ} = \frac{I_{CQ}}{h_{FE}} \text{ (refer to BC108 data for } h_{fe} \text{ value)}$$

$h_{FE} = 110 - 800$ − Take the lowest value

$$I_{BQ} = \frac{10.7\text{mA}}{110} = 97.2\mu A$$

$$I_{BQ} = 97.2\mu A$$

It is now time to determine the base bias resistors R_1 and R_2 that will fix the base current and the base voltage V_B.

9 Find the base voltage V_B.

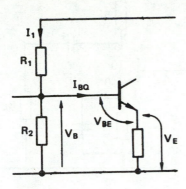

$$V_B = V_E + V_{BE}$$
$$V_E = 1.07V$$
$$V_{BE} \simeq 0.6V \text{ (for a Si transistor)}$$
$$\therefore V_B = 1.07 + 0.6$$
$$= 1.67V$$

10 Decide on a value for I_1.

I_1 is the current through the potential divider R_1, R_2.

Rule of thumb – I_1 should be greater than ten times I_{BQ}.

$$I_{BQ} = 97.2\mu A$$
$$I_1 = 97.2\mu A \times 10 = 972\mu A$$
$$\therefore \text{ Make } I_1 = 1mA$$

11 Calculate value of R_2.

$$V_B = 1.67V$$
$$I_1 = 1mA$$
$$\therefore R_2 = \frac{1.67V}{1mA} = 1.67k$$
$$\text{n.p.v.} = 1.5k$$
$$\therefore R_2 = 1.5k$$

12 Calculate value of R_1.

$$R_1 = \frac{V_{CC} - V_B}{1mA} = \frac{12 - 1.67}{1mA}$$
$$= \frac{10.3}{1mA} = 10.3k$$
$$\text{n.p.v.} = 10k$$
$$\therefore R_1 = 10k$$

13 Decide on values for coupling capacitors. The lowest signal frequency is 500Hz, so capacitive reactance must be low at 500Hz, e.g. $10\mu F$;

$$X_C = \frac{1}{2\pi f C} \ \Omega$$
$$= \frac{1}{6.28 \times 500 \times 10 \times 10^{-6}}$$
$$= 31.8\Omega$$

This acts as a series resistance.

$$\therefore C_1, C_2 = 10\mu F$$

14 Decide on a value for decoupling capacitor C_E. To preserve the low frequency gain the reactance of C_E at the lowest signal frequency must be very low, e.g. at 500Hz;

$$100\mu F = \frac{1}{2\pi(100 \times 10^{-6} \times 500)} = 3.18\Omega$$

This is a much lower resistance than R_E and will therefore be suitable.

$$\therefore C_E = 100\mu F$$

15 Draw the designed circuit complete with values.

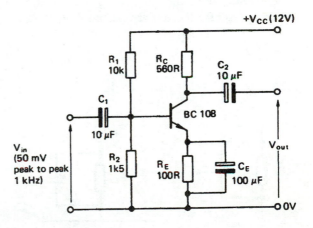

16 Now check on the gain of the circuit using the transistor model.

$$\delta V_{CE} = 6V$$
$$\therefore V_{ce} = 6V \text{ (lower case indicates signal voltage)}$$

The current change δI_C will be

$$\frac{V_{ce}}{560} = 10.7mA$$

So I_C will increase by 5.35mA and decrease by 5.35mA about its Q point value of 10.7mA

$$I_c = 10.7\text{mA}$$
$$gm = \frac{I_c}{25} = \frac{10.7}{25} = 428\text{mS}$$
$$A_v = -gm \times R_c$$
$$= 428 \times 10^{-3} \times 560 = -240$$
$$\therefore A_v = -240$$

This is twice the specified gain of 120.

But before changing any values build and test the circuit then make adjustments.

Building and testing the circuit

1 Connect the circuit and check your d.c. bias voltages by measuring V_{CE}, V_B and V_E with respect to the 0V line, and comparing these with your calculated values.

2 If this checks out, the gain can be adjusted by modifying the amount of a.c. negative feedback. Remember C_E prevents any a.c. negative feedback (it prevents the signal flowing through R_E). If R_E is split then some signal feedback will occur, and hence the gain will be reduced. Use a preset for R_{E1} and 'tweak' until $A_v = 120$.

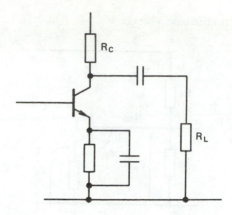

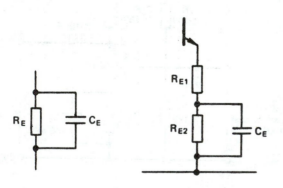

3 By altering the value of R_C, A_v can be changed.

4 *Loading*

The gain will be affected by anything connected to your circuit. This is because any resistor, component or piece of equipment connected across the output terminal can be considered as a load R_L.

It can be shown that the value of R_L from a signal point of view will act in parallel with R_C

and change the gain of the amplifier, so you can see that even the oscilloscope you use to measure the output will slightly affect the gain.

5 By introducing an actual R_L into your circuit you will in fact change the gain, i.e. if an R_L of 560Ω were included:

$$A_v = -gm \times R_C \| R_L$$
$$(\| \text{ means in parallel with})$$
$$= -428 \times 10^{-3} \times 280 = -120$$

All these factors can only be determined by the building and testing aspect of design. Once the testing and modifications are complete a test report can be written.

Writing a test report

Having designed, built and tested your amplifier it makes sense to have a fully documented account of the exercise. This should include the following information.

1 The design specification, i.e. what is expected of the finished circuit.

2 The proposed circuit together with the calculations showing how each resistor value has been decided. Explain any assumptions made and reasons for selecting the appropriate transistors.

3 Details of the testing procedure, this must include

a) the gain at the specified frequency,
b) signal distortion level,
c) gain–frequency response curve,
d) bandwidth,
e) input and output resistance.

DESIGN ASSIGNMENT 5

Amplifier (2)

A common source Class A audio amplifier is to be designed that must meet the following specifications.

1 Device – n channel JFET
2 Power supply – 10V d.c.
3 Voltage gain – 8 when V_{in} is a 1kHz sine wave of 100mV pk–pk.

4 Bandwidth – 10Hz–20kHz.

In addition to the above specifications a full test report will be required that includes all information relating to the following.

a) d.c. voltage levels,
b) signal distortion levels,

---

4 Details of any circuit modifications that were required to bring the amplifier up to specification.
5 The complete modified circuit diagram.
6 An appraisal of the circuit, highlighting any shortcomings in the circuit and problems that have been encountered.

This may seem quite a massive undertaking but the reality is that all of the work has been done during the designing, building and testing stages. All the documentation does is present your labours in a professional manner that will serve as a record and reference document for all time.

Designing FET amplifiers

THE JFET COMMON SOURCE AMPLIFIER

This device operates in depletion mode. Therefore the Q point must be established by making the gate source voltage (V_{GS}) negative in the case of an n channel device, or positive in the case of a p channel device.

The voltage gain can be approximated using $A_v = -gm\,R_D$, where gm is the transconductance value. A glance at the data sheets at the quoted values for gm (remember this may appear as Y_{fs} or gfs) will show that these values tend to be low for JFETs, e.g. for a 2N3819, Y_{fs} (min) is 2000μS. So if an amplifier has a drain resistor R_D of 5k the voltage gain would be

$$A_v = -2000 \times 10^{-6} \times 5000$$
$$= -10$$

So a point to bear in mind is that the common source amplifier will probably have a lower voltage gain than its common emitter relation.

Procedure

Starting point
As always, sketch the proposed circuit and deal with the known facts.

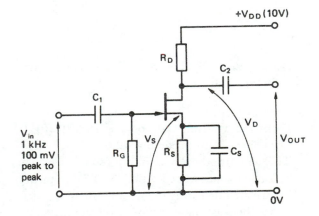

1 Calculate V_{out}.

$$A_v = \frac{V_{out}}{V_{in}}$$
$$V_{in} = 100\text{mV}$$
$$A_v = 8$$
$$\therefore V_{out} = V_{in} \times A_v$$
$$= 100\text{mV} \times 8 = 800\text{mV}$$
$$V_{out} = 800\text{mV pk–pk}$$

This is a very small swing of ± 400mV about the Q point. It is unlikely to produce distortion, so although class A biasing is specified you can see that anywhere on the load line will probably be all right in this case.

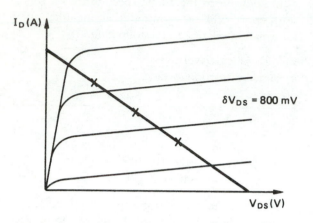

2 Select a device.

Just as you needed to know the quiescent value of base current (I_{BQ}) for the BJT you will require a $-V_{GSQ}$ in order to fix the Q point. This value will vary with the chosen device so first select a suitable transistor with reference to the required specifications:

$$I_D \text{ (max)} = 10\text{mA}$$
$$+V_{DD} = 10\text{V}$$

Almost any general purpose FET will do. I have a 2N3819 available.

3 Determine a value of $-V_{GS}$.

From the available data on the 2N3819 (FET Data Sheet 2) the maximum value of $V_{GS} = 8$V. This will turn the device fully off so it is sensible to look at a quiescent V_{GS} of about 3V (remember we are calculating theoretical values in order to use them as a base to work from in practice).

$$V_{GSQ} = -3.0\text{V}$$

4 Determine R_D.

This must now be found because the drain resistor R_D affects the gain and the drain current I_D.

$$A_v = 8$$
$$gm = 2\text{mS (min)}$$

Since
$$A_v = -gm R_D$$
$$R_D = \frac{A_v}{gm} = \frac{8}{2 \times 10^{-3}} = 4000\Omega$$
$$\text{n.p.v.} = 3.9\text{k}\Omega$$
$$\therefore R_D = 3.9\text{k}$$

5 Calculate I_D.

$$V_D \simeq \frac{V_{DD}}{2} = 5\text{V (Class A bias)}$$
$$\therefore I_D R_D = V_{DD} - V_D$$
$$= 10 - 5 = 5\text{V}$$
$$\text{If} \quad R_D = 3.9\text{k}$$
$$I_{DQ} = \frac{5.0}{3.9\text{k}} = 1.28\text{mA}$$

6 Calculate R_s.

$$I_S R_S = 3\text{V (because the source resistor } R_S$$
$$\text{effectively biases the gate)}$$

So, if $I_S = I_D = 1.28$mA

$$R_s = \frac{3\text{V}}{1.28\text{mA}} = 2.34\text{k}$$
$$\text{n.p.v.} = 2.2\text{k}$$
$$\therefore R_S = 2.2\text{k}$$

7 Determine R_G.
R_G simply gives the gate a 0V reference level in order that the gate source biasing is correct. It can be any high value, typically $1\text{M}\Omega$.

$$R_G = 1\text{M}$$

8 Determine C_{in}, C_{out}, C_S.
As with the common emitter amplifier, these should have a low reactance at signal frequencies. 10μF will be suitable for C_{in} and C_{out}, with 100μF for decoupling capacitor C_S.

9 Draw the theoretical circuit overleaf complete with values.

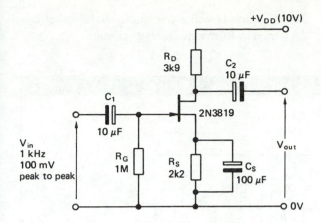

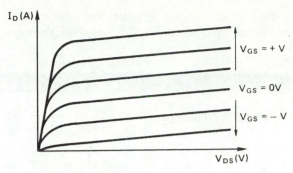

Fig. 3.46 n channel depletion type MOSFET

Now build the circuit and check the gain at 1 kHz with a V_{in} of 100mV.

What if it's out?

Make adjustments – but not in a random manner! Work in a logical fashion using your knowledge about how the circuit works. Due to the resistance of the device itself (R_{ds}) a voltage will be dropped across the actual transistor. Consequently R_S can only be approximate and may have to be altered to achieve the required A_v.

Remember Whatever is connected to your amplifier output will change the overall loading such that the gain

$$A_v = -gm \times R_D \| R_L$$

where R_L = load connected to the amplifier (‖ means in parallel with).

Now continue with the full amplifier test and the compilation of the test report.

MOSFET amplifier design

DEPLETION TYPE

You will remember that a depletion-type MOSFET will operate in both enhancement and depletion mode (Fig. 3.46). This means that the device can actually be biased with the gate at 0V if required. V_{GS} will swing positive and negative with the input signal to produce δV_{GS} (V_{gs}).

ENHANCEMENT TYPE

This device will only operate in enhancement mode and must be biased accordingly (Fig. 3.47).

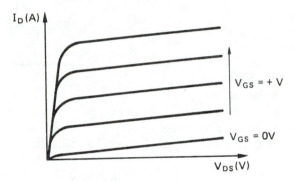

Fig. 3.47 n channel enhancement type MOSFET

CIRCUIT ARRANGEMENTS

The MOSFET can be biased using the same techniques employed for the BJT and JFET with a typical circuit as shown in Fig. 3.48.

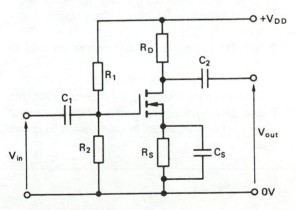

Fig. 3.48 MOSFET amplifier circuit

Now that you are familiar with the various individual types of amplifier investigation, Practical Investigation 18 will allow you to see how they can be joined to form a multi-stage circuit.

Two stage amplifier

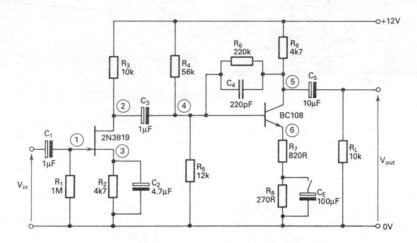

Equipment
BC108 transistor
2N3819 field effect transistor
270R, 820R, 1k, 2 × 4k7, 2 × 10k, 12k, 56k, 220k, 1M resistors
220pF, 2 × 1μF, 4.7μF, 10μF, 100μF capacitors
Dual beam oscilloscope (CRO)
Power supply
Signal generator
6 cycle log/lin graph paper
Digital multimeter

Method
1 Build the circuit shown above, ensuring that there is no a.c. negative feedback applied (stage 2 bypass capacitor C_E is *in* circuit).
2 Check that the circuit is operational, measure and record the following:
   a) the input signal distortion level at 1kHz
   b) the voltage gain of each stage and the overall circuit at 1kHz.
3 Under quiescent conditions (no signal!) measure the voltages at points 1–6 in the amplifier and the current drawn by the complete circuit.

4 By taking measurements at appropriate steps produce a frequency response curve (voltage gain against frequency) for the amplifier.
5 Introduce a.c. negative feedback by switching capacitor C_E *out*. Repeat Step 4 plotting the graph on the same axis.
6 Switch C_E back in circuit and set the input signal to 1kHz. Note what happens if RL is
   a) removed,
   b) changed to 1k.

Results
1 Using your graph determine the bandwidth of the amplifier for the case with and without negative feedback applied.
2 Make an observation about the relationship between negative feedback and voltage gain.
3 What effect does the load that the amplifier has to drive have on its voltage gain?
4 Give full details of the methods by which a.c. negative feedback is applied in this circuit.

The results of this assignment should be presented in the form of a comprehensive report that includes all the specifications and recorded data.

Amplifier review

- An active device (BJT or FET) can be used to provide voltage current or power gain.
- Gain is a measure of amplification given the symbol *A*.

$$A_v = \text{voltage gain} = \frac{V_{out}}{V_{in}}$$

$$A_i = \text{current gain} = \frac{I_{out}}{I_{in}}$$

$$A_p = \text{power gain} = \frac{P_{out}}{P_{in}}$$

- A d.c. load line can be used to fix the quiescent bias point (Q point) for a transistor.
- Quiescent means 'under no signal conditions'.
- For a class A amplifier the Q point is fixed about half way along the load line.
- A common emitter and a common source amplifier are inverting amplifiers.
- Capacitors are used to allow a.c. signals to pass whilst blocking d.c.
- A BJT comprises 'p' and 'n' type semiconductor materials, each having minority carriers present.
- Minority carriers cause a leakage current across a reverse bias junction that increases with temperature. Therefore BJTs require thermal stability.
- Feedback occurs when a fraction of an amplifier's output signal is fed back to the input.
- Positive feedback occurs when the feedback signal is 'in phase' (zero phase shift) with the input signal.
- Negative feedback occurs when the feedback is in 'anti-phase' (180° phase shift) with the input signal.
- Positive feedback produces instability in an amplifier and should be avoided.
- Negative feedback affects the performance of an amplifier in the following ways.
  - **a)** The gain of the amplifier will decrease.
  - **b)** The bandwidth will increase.
  - **c)** The gain stability will improve.
  - **d)** The noise and distortion will reduce.
  - **e)** The input and output impedances will change.
- Feedback can be a.c. or d.c.
- A bypass or decoupling capacitor is included across the emitter and source resistor to prevent a.c. signals passing through the resistor (negative feedback).
- Removal of the bypass capacitor will cause a.c. negative feedback to occur and significantly reduce the gain of the amplifier.
- When considering an amplifier under signal conditions, lower case subscripts are used to denote 'small signals'.
- An 'n channel' FET requires the gate to be biased negative with respect to the source; this is achieved by the source resistor R_S.
- An amplifier can be represented by a 'small signal model' where the input to the device is represented by a *Thevenin* equivalent circuit and the output represented by a *Norton* equivalent circuit.
- The small signal models of BJT and FET amplifiers allow the voltage gain to be determined theoretically.
- An amplifier can be designed to have a narrow bandwidth giving a high gain at a specific frequency. This is achieved by replacing the load resistor with a tuned circuit.
- A current amplifier can be used to act as a buffer between a circuit and the load connected to it.
- When designing an amplifier always work from the known to the unknown.
- Be careful when deciding on the specifications if these are not already given; *do not give yourself an impossible task!*
- If the required specifications are given analyse them carefully – it helps to write a list of the salient points.
- When calculating values, look for answers that are obviously out, e.g. 0.01 Ω for R_C or 2700 MΩ for R_1.
- Use preferred values for components and round up/down your results accordingly. Then check that the change will not affect the result too much.
- When the circuit is tested, if the results are not what you expected check everything *before* making changes.
- Make changes in a logical manner and think them through.
- When all changes have been made remember to enter these in the documentation.

SELF ASSESSMENT ANSWERS

Self assessment 7

1 $I_{CQ} = 2mA$, $R_C = 2k2$, $V_{CC} = +12V$

$$\begin{aligned}
V_{CEQ} &= V_{CC} - (I_C R_C) \\
&= 12 - (2 \times 10^{-3} \times 2.2 \times 10^3) \\
&= 12 - 4.4 = 7.6V
\end{aligned}$$

2 V_{CE} can move from 7.6 to 0V but only from 7.6 to +12V when clipping will occur. Therefore maximum output swing is from 3.2V to 12 V = $\pm 4.4V$

3 $I_{CQ} = 2mA$ when $V_{CEQ} = 7.6V$
$I_{CQ}R_C = 4.4V$ because $V_{CC} - V_{CE} = I_C R_C$
When V_{CE} swings to 12V
$I_C R_C = 12 - 12 = 0V$

When V_{CE} swings to 3.2V
$I_C R_C = 12 - 3.2 = 8.8V$
If $I_C R_C = 8.8V$ and $R_C = 2k2$

then $I_C = \dfrac{8.8V}{2.2 \times 10^3} = 4mA$

The maximum swing of I_C before distortion occurs is $\pm 2mA$.

4 $I_{CQ} = 2mA$, $h_{FE} = 120$

$$I_B = \frac{I_C}{h_{FE}} = \frac{2 \times 10^{-3}}{120} = 16.6\mu A$$

Multiple choice questions

1 An amplifier has a gain (A_p) of 520. If the input signal is 85mW the output signal will be:
a) 6.1kW
b) 442mW
c) 44.2W
d) 4.4kW

2 A transistor is to be used as a class A signal amplifier. If the supply voltage is to be 12.0V d.c. a suitable quiescent output voltage (V_{CEQ}) would be:
a) 3.0V
b) 0.6V
c) 9.0V
d) 6.0V

3 With reference to the small signal amplifier shown in Fig. 3.49.

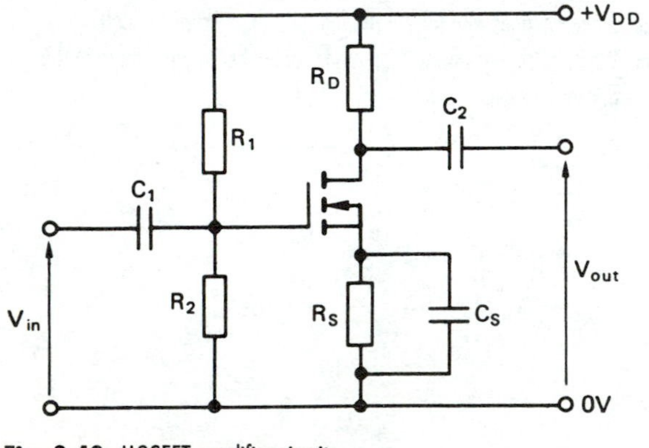

Fig. 3.49 MOSFET amplifier circuit

The purpose of capacitor C_1 is to
a) provide a.c. negative feedback
b) provide thermal stability
c) provide a.c. coupling
d) provide d.c. feedback

4 With reference to Fig. 3.49. Capacitor C_S is included in the circuit to:
a) provide a.c. biasing for the transistor
b) provide thermal stability
c) prevent a.c. negative feedback
d) prevent d.c. negative feedback

5 If negative feedback is applied to a signal amplifier select the option that accurately describes the result.
a) The gain increases but the bandwidth decreases.
b) The gain decreases and the bandwidth increases.
c) The gain and bandwidth remain the same.
d) The gain remains unchanged but the bandwidth increases.

6 For negative feedback to occur in an amplifier there must be a phase shift from output to input of:
a) 180°
b) 90°
c) 360°
d) −180°

7 A transistor used as an amplifier in common emitter mode has a collector load resistor of 4k7Ω. If the collector current (δI_C) under signal conditions is 4mA the voltage gain A_v will be:

a) -85 c) -188

b) -851 d) -752

8 A junction field effect transistor is to be used as a class A common source amplifier. If the *gm* is quoted as -15mS and a drain load resistor of 6k8Ω is used the voltage gain will be:

a) 102   c) 22.05

b) 0.45   d) -102

9 A tuned voltage amplifier will have:

a) a voltage gain that is constant over a wide frequency range

b) a very wide bandwidth

c) a low gain and wide bandwidth

d) a narrow bandwidth

10 A current amplifier will:

a) provide phase inversion

b) have a voltage gain of 1

c) have a high voltage gain

d) have identical input and output resistances

STABILISED POWER SUPPLIES

The purpose of a stabilised or regulated power supply is to keep the d.c. output constant despite variations in the a.c. supply voltage and the d.c. load current.

A very good power source for electronic circuits is, of course, a battery, and indeed this is the only supply used for portable equipment. The disadvantage is the cost, so where possible it makes good sense to use the readily available mains a.c. supply and convert it to the required d.c. A rectified and smoothed power supply circuit was considered in Chapter 1: although this is suitable for a number of applications, the output it provides is not smooth or stable enough for powering electronic circuits.

A good starting point is to consider the power supply as a unit comprising a series of blocks as shown in Fig. 4.1.

THE TRANSFORMER

This takes the mains supply voltage and steps it down to the required voltage, e.g. 240V a.c. to 12V a.c. There are, of course, occasions when the mains supply voltage is too low in which case a step up transformer would be used.

THE RECTIFIER

This is the part of the circuit that converts the a.c. voltage into a unidirectional or d.c. voltage. Half-wave rectifiers are not used in stabilised power supplies so the rectifier block would contain a full-wave rectifier. The bridge type of full-wave rectifier was fully discussed in Chapter 1 and this is the version most often found in use today.

However, an alternative circuit shown in Fig. 4.2 employs a 'centre tapped' transformer. Study the waveforms shown in Fig. 4.3. The centre tap (B) is taken as the point of reference. With respect to point B the voltages at A and C are in anti-phase (phase shifted by 180°). When A is positive with respect to B diode D_1 conducts and current flows through the load R_L developing a voltage across it. When C is positive with respect to B diode D_2 conducts and voltage is developed across R_L in the same way.

You may see from the waveforms that only one diode conducts at any time, the other is reverse biased. The result of this is that current flows in the load resistor for the full cycle of a.c. voltage and the output voltage is identical to that obtained from the bridge circuit previously discussed.

THOUGHT

■ *The centre tapped transformer circuit uses only two diodes, the bridge circuit uses four. Why is the bridge circuit more popular since it uses more components?*
A centre tapped transformer is costlier than one with a simple secondary winding.

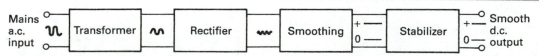

Mains a.c. input — Transformer — Rectifier — Smoothing — Stabilizer — Smooth d.c. output

Fig. 4.1 Block diagram of stabilised power supply

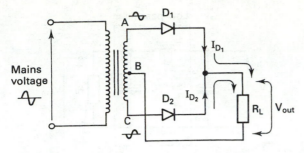

Fig. 4.2 Full-wave rectifier using a centre tapped transformer

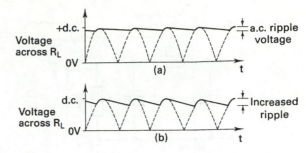

Fig. 4.4 (a) a.c. ripple voltage. (b) Increased ripple

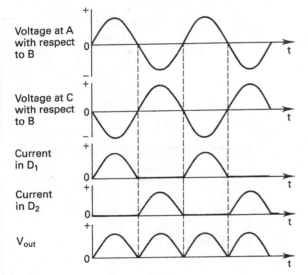

Fig. 4.3 Full-wave rectifier waveform

Smoothing

The unidirectional voltage produced by the full-wave rectifiers consists of a series of positive half cycles occurring at a frequency of 100Hz. As discussed in Chapter 1, a reservoir capacitor converts this 'raw' d.c. into a smoother version but this will have some a.c. ripple present as shown in Fig. 4.4a. The size of the ripple voltage is determined by the current that is drawn from the supply, if the current demand increases the ripple voltage will also increase (Fig. 4.4b).

The d.c. can be considered as the low frequency part of the signal (d.c. = 0Hz) and the a.c. ripple is the high frequency part of the signal (ripple = 100Hz). To reduce the ripple a low-pass filter can be included that allows the d.c. to pass

while attenuating (reducing) the size of the ripple voltage.

There are two types of low-pass filters used.

THE RC FILTER

Fig. 4.5a shows a circuit. Note that the reservoir capacitor C_1 is used as the first line in smoothing. The resistor, capacitor (RC) network reduces the ripple in the following way.

Capacitor C_2 will not pass d.c. so the circuit can be redrawn (Fig. 4.5b) to show how it affects the d.c. part of the signal. R_1 and R_L make a potential divider circuit for the voltage stored in C_1 (V_{in}). Provided R_1 is low compared with R_L the d.c. output voltage (V_{out}) will be almost the same as V_{in}.

Fig. 4.5c now considers the circuit from the a.c. ripple point of view. The a.c. ripple current that would normally flow through the load resistor R_L is now bypassed via C_2, consequently the ripple voltage present in the output will be reduced.

THOUGHT

■ *Why won't the ripple voltage be completely removed?*
The reactance of the bypass capacitor C_2 will determine how much the ripple will be reduced. If the reactance of C_2 is very low compared with the resistance of R_L then not much ripple voltage will be developed across R_L. The actual reactance (X_{C2}) can be calculated using:

$$X_{C2} = \frac{1}{2\pi f C_2} \ \Omega \ \text{where } f = \text{ripple frequency}$$

Note   However low the reactance of C_2 is there will always be some ripple developed across R_L.

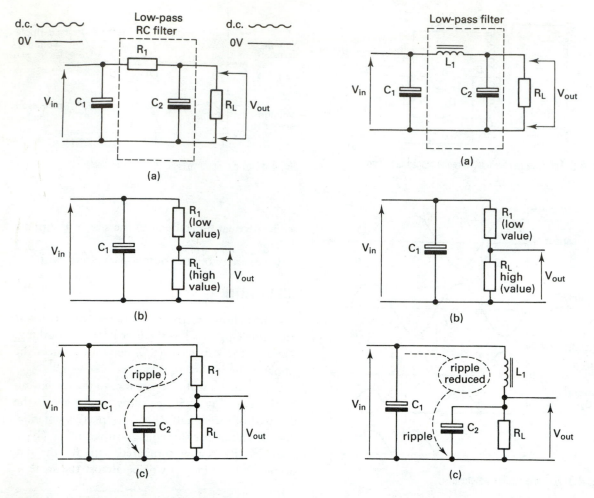

Fig. 4.5 (a) RC low-pass filter network. (b) Equivalent circuit for d.c. signal. (c) Equivalent circuit for a.c. signal

Fig. 4.6 (a) CL low-pass filter. (b) Equivalent circuit for d.c. signal. (c) Equivalent circuit for a.c. signal

The CL filter

Fig. 4.6a shows the circuit with C_1 as the reservoir capacitor. Here a capacitor and inductor provide the ripple reducing filter which operates in the following way.

The inductor (L_1) will have a d.c. resistance R_1. Fig. 4.6b shows the circuit redrawn so that its behaviour to d.c. signals can be analysed. C_2 is open circuit to d.c. and so is not included. If R_1 is low in value compared with the load resistor R_L the majority of the input voltage V_{in} will appear across the load as V_{out}. From the a.c. ripple point of view the circuit can be represented by that

shown in Fig. 4.6c. Here the inductor L_1 offers a high reactance to the a.c. ripple since:

$$X_L = 2\pi f L \; \Omega \quad \text{where } f = \text{ripple frequency}$$

Capacitor C_2 however offers a low reactance to the a.c. ripple since:

$$X_{C_2} = \frac{1}{2\pi f C_2} \; \Omega$$

The reactance of L_1 is chosen to be large compared with the reactance of C_2 with the net result that a much reduced ripple appears across the load resistor R_L.

This type of low-pass filter is more efficient than the RC type. However it tends to be used for high voltage d.c. power supplies, e.g. 100V and above.

THOUGHT

■ *If the CL filter is better why is it not used for low voltage supplies? For the inductor to have the high reactance required to reduce the ripple it must have a large number of turns. This means that the d.c. resistance of the inductor will be high enough to drop a significant d.c. voltage across it. Inductors also tend to be costly and bulky.*

Stabilisers

All power supplies have a 'no load voltage'; this is the output voltage without any load current being drawn from the supply. Let's say it is 12V d.c. Since no current is being drawn this output voltage is likely to be extremely smooth and stable, i.e. it will be free from 'ripple' and not vary at all (Fig. 4.7).

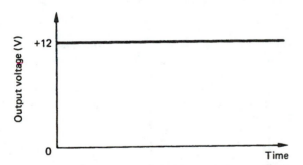

Fig. 4.7 Smooth stable power supply output

If a load is now connected to the circuit that draws 100mA the output voltage may:

a) fall slightly,
b) develop 'ripple',
c) do both.

The greater the output current the more the output voltage will be affected (Fig. 4.8).

Even with no load connected, variations in mains supply must be considered. Although the supply is nominally 240V, 50Hz it can vary about

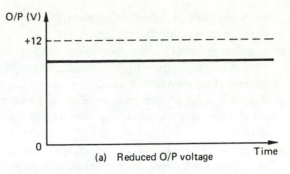

(a) Reduced O/P voltage

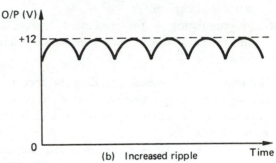

(b) Increased ripple

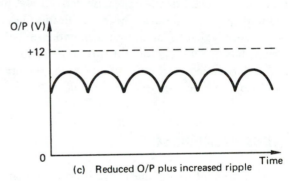

(c) Reduced O/P plus increased ripple

Fig. 4.8 (a–c) Changes in output for a non-stabilised output

the quoted value and this change will cause the output voltage to increase and decrease.

A stabiliser circuit must therefore ensure that even though the mains voltage may vary slightly and the output current changes, the d.c. output voltage remains as constant and as smooth as the application demands. This means that regulated power supplies are built to a specification that will include the following parameters.

■ Output voltage – maximum and minimum limits of output d.c. voltage, e.g. variable 0–30V or 12V fixed.
■ Output current – maximum current that can be drawn from the power supply before it ceases to

be within specification or protects itself from damage by limiting the current.

■ Load regulation – the maximum change in output voltage due to a change in output current from zero to full load.

■ Ripple and noise – the amount of a.c. appearing on the d.c. output, usually quoted at full load current.

■ Stability – the change in d.c. output voltage with time for a fixed value of input voltage and output load current.

■ Line regulation – the maximum change in output voltage due to a change in a.c. input voltage.

It follows that the more demanding the specification, the more complicated will be the circuit and consequently the more expensive it will be to produce and buy.

Generally, it is the output current requirement that dictates cost. For example, it is relatively cheap to build a very good quality high specification 12V power supply that has an output current demand of 50mA, but one with an output current demand of 25A will be much more costly if the ripple, stability and load regulation are to be reasonable.

Types of stabilisers

We shall be considering linear stabilisers that are in common usage, although it is worth mentioning that where high powers are required the *switched mode power unit* (SMPU) is finding widespread application.

THE SHUNT STABILISER

This has been dealt with in Chapter 1. The Zener diode stabiliser (Fig. 4.9) was considered – briefly to recap.

The total current (I_t) is shared between the Zener and the load such that

$$I_t = I_z + I_L$$

Any increase or decrease in I_L produces changes in Zener current so that the output voltage remains constant. This circuit has a number of points that make it quite unsuitable for anything other than basic regulation:

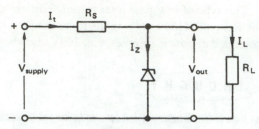

Fig. 4.9 Simple Zener diode shunt stabiliser

1 The output voltage is the Zener diode voltage.
2 There is no way of adjusting the output voltage.
3 The Zener diode dictates the output current, i.e. if you want a 5A output you need a 5A Zener.
4 There is no monitoring of output current demand so the circuit is relatively insensitive.

An improvement is to use a power transistor to supply the current and use the Zener diode as a voltage reference source (Fig. 4.10). In this circuit TR_1 is used as a current amplifier (emitter follower). The Zener diode holds the base at a constant voltage, e.g. if the Zener diode has a voltage rating of 8.2V then the output voltage of the supply will be approximately 7.6V (0.6 V_{BE} for TR_1).

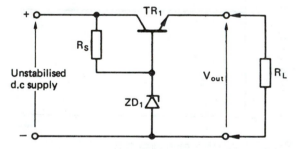

Fig. 4.10 Zener diode used as a voltage reference source

This circuit, although an improvement, is really only a method of reducing the Zener power requirement by sharing it with a power transistor. There are still the disadvantages of (a) quite high Zener powers if the output current is high and (b) limited sensitivity.

To overcome these shortcomings, a second transistor can be used to monitor the output voltage and, if this changes, adjust the output current to restore it.

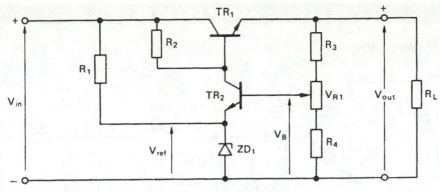

Fig. 4.11 Basic circuit for most series power supplies

Fig. 4.11 shows the basic circuit for most series power supplies with the components having the following functions.

1 ZD_1 is a Zener diode that provides a reference voltage (V_{ref}) that holds the emitter of TR_2 at a fixed potential (the Zener voltage).
2 R_1 is the series Zener diode resistor.
3 R_2 is the base bias resistor of TR_1 and the collector load resistor for TR_2. It is the voltage across this resistor that will determine how much TR_1 will conduct.
4 TR_1 is the series transistor that controls the output current of the power supply and hence the output voltage.
5 TR_2 is the transistor that monitors the output voltage, sometimes called the *error amplifier* because it detects any error in V_{out}, amplifies it and uses this signal to restore the output.
6 R_3, V_{R_1}, R_4 make up a variable potential divider that will allow the output voltage (V_{out}) to be adjusted using V_{R_1}.

In practice the circuit works in the following way: let's imagine that V_{out} is 10.0V d.c.:

1 TR_2 emitter is held at a constant voltage (V_{Ref}) by the Zener diode.
2 $V_B = V_{Ref} + 0.6$ (V_{BE} of TR_2).

This voltage V_B determines the collector current of TR_2 and hence the voltage drop across R_2. Now the voltage across the resistor R_2 is the *base bias* voltage of the series transistor TR_1 and this determines the output voltage.

Consider an increase in load current: the voltage across TR_1 increases so the output voltage V_{out}

falls. To restore V_{out} the stabiliser must compensate by increasing the current flow through TR_1, this is achieved in the following way.

The fall in V_{out} results in a reduction in V_B. Since TR_2 emitter is held constant by V_{Ref}, the forward bias of TR_2 is reduced (because $V_B - V_{Ref}$ is lower). TR_2 now conducts less, so the voltage drop across R_2 is reduced and the base bias voltage of TR_1 increases, thus supplying a higher output current restoring V_{out} to its original value of 10.0V.

If the output voltage increases due to a rise in V_{in}, V_B goes up so ($V_B - V_{Ref}$) increases and TR_1 conducts more. Even more current flows through R_2 so the voltage drop across it increases and the base bias of TR_1 is reduced, making it conduct less, and V_{out} is restored.

Points for consideration with this circuit

1 For a high current power supply TR_1 will have to be a power transistor mounted on a heat sink.
2 If you look at the transistor data you will see that the higher the power of the transistor the lower its h_{FE}, e.g. a 2N3055 is capable of supplying 15A but its h_{FE} is only 20–70.
3 TR_2 supplies the base current for TR_1. If TR_1 is a 2N3055 supplying 10A with an h_{FE} of 20, then TR_2 must supply

$$\frac{10A}{20} = 500mA$$

So, TR_2 must itself supply a fairly high current and have a high gain. For this reason high

PRACTICAL INVESTIGATION 19

The stabilised power supply

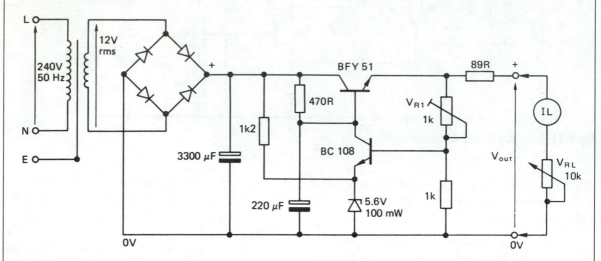

Objective
To build and test a stabilised power supply and produce a full specification of the unit in the form of a comprehensive test report.

Method
1 Build the circuit (as the output current is only 100mA there is no need for a heat sink for the BFY51).
2 With RL disconnected, measure and record the no-load output voltage range as determined by V_{R1}.
3 Starting with RL set to 10kΩ and reducing in appropriate steps, record V_{out}, IL and the ripple voltage for each step until IL = 100mA.
4 Sketch the output voltage waveform under full load conditions indicating the ripple and d.c. level.
5 Plot a graph of V_{out} against IL over the operating range.
6 Determine the load regulation using:

$$\text{load regulation} = \frac{\text{no load voltage } - \text{ full load voltage}}{\text{no load voltage}} \times 100\%$$

7 Calculate the output resistance of the power supply using:

$$R_{out} = \frac{\delta V_{out}}{\delta IL}\ \Omega$$

Results
1 Use the results obtained from the investigation to produce detailed specifications for the power supply relating to:
■ the output voltage
■ the load current
■ the regulation obtainable over the operating range
■ the output resistance of the supply
■ the maximum ripple voltage
For each of these be sure to define the terms and explain what they mean in practice.
2 With reference to the circuit, identify the constituent circuits that perform the following functions:
■ rectification
■ smoothing
■ regulation
Full details of how these operate should be provided.

current power supplies use a modified version of this circuit that employs further amplification.

Practical Investigation 19 will enable you to become familiar with a regulated power supply and test its operation for yourself.

INTEGRATED CIRCUIT VOLTAGE REGULATORS

Today the design of power supplies has been greatly simplified by the fact that there are now a number of voltage regulator integrated circuits available, so most of the required electronics are

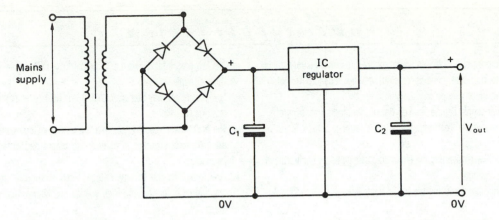

Fig. 4.12 Power supply using IC regulator

contained in a single package reducing the power supply circuit diagram to that of Fig. 4.12.

The benefits of integrated circuit regulators are:

1 they are available in a comprehensive off-the-shelf range of voltages and currents
2 they have inbuilt protection from overloads and short circuits
3 they are very easy to connect in circuit
4 variable output ICs are available.

Designing a power supply using an IC regulator

The only real problem facing the designer here is which device to select. If you scan the data for IC regulators, you will see that the choice can be made by considering the following questions.

1 Does the supply require a fixed or variable output?

2 What output voltage or voltage range is required?
3 What is the maximum current required?

SELF ASSESSMENT 8

Consider the following power supply specifications and select from the data a suitable IC for each application.

1 +5V, 50mA.
2 1.0A supply with V_{out} variable 2–30V.
3 −12V, 800mA.
4 9A, 10V.
5 350mA, variable 10–32V.

A point to remember is that the higher the current required the more expensive the device is likely to be.

SELF ASSESSMENT ANSWERS

Self Assessment 8
1 Fixed regulators above 50mA, e.g. LM342 p-5, LM78L05ACH, LM78M05CT etc.
2 Adjustable LM317F, LM337T etc.
3 Fixed negative above 800mA, e.g. L7912CV, LM320T−12 etc.

4 No fixed supply available from the given range at 9.0A. Choose variable LM396k with 10A capability.
5 Variable above 350mA, e.g. LM317MP, LM317H etc.

Power supply review

- The purpose of a stabilised power supply is to provide a very stable smooth d.c. that remains constant despite variations in output current, input voltage and temperature.
- A power supply can be 'broken down' into blocks consisting of:
  the transformer (converting mains a.c. voltage into a lower or higher voltage)
  the rectifier (converting the alternating voltage into a unidirectional or d.c. voltage)
  the smoother (reducing the mains ripple voltage to an acceptable level)
  the stabiliser or regulator (that keeps the d.c. output voltage constant)
- The smoothing unit will consist of a reservoir capacitor plus a low-pass filter.

- The low pass filter allows d.c. to pass (d.c. = 0Hz) while reducing the a.c. ripple voltage.
- Low-pass filters may be resistor, capacitor (RC) types or capacitor, inductor (CL) types.
- The best power supplies maintain the output voltage constant by using an error amplifier to monitor the output and correct when necessary.
- Power supplies can be any voltage, fixed or variable, supplying any desired output current. High voltage and high current supplies are costly.
- Integrated circuit regulators simplify power supply design and construction and are widely available.

Multiple choice questions

1 With reference to the d.c. power supply shown in Fig. 4.13. Block 2 would consist of?
a) a smoothing capacitor
b) a rectifier
c) a stabiliser
d) a primary filter

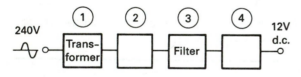

Fig. 4.13

2 With reference to the diagram of Fig. 4.13. Block 4 would be?
a) a rectifier
b) a filter
c) a smoothing capacitor
d) a stabiliser

3 A stabilised mains power supply uses a centre-tapped transformer and two diodes to provide the raw d.c. for the circuit. The a.c. ripple voltage present at the output would have a frequency of:
a) 50Hz
b) 100Hz
c) 25Hz
d) 0Hz

4 A power supply has a no-load voltage of 15.0V d.c. This falls to 14.96V when the supply is delivering maximum load current of 850mA. The load regulation would be:
a) 2.6%
b) 4%
c) 0.34%
d) 0.26%

5 The power supply referred to in Question 4 will have an output resistance of:
a) 17.6Ω
b) 47Ω
c) 0.047Ω
d) 1.76Ω

5

COMBINATIONAL LOGIC

Analogue signals

An analogue signal is a voltage or current that varies smoothly between two extremes. Consider the waveform shown in Fig. 5.1.

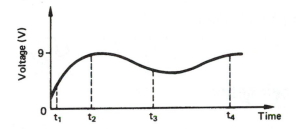

Fig. 5.1 Analogue voltage signal

At any instant along the time axis, the voltage will have a certain value so it is easy to see that there is an infinite number of possible levels between 0V and 9V! Many of the circuits considered have been analogue, e.g. amplifiers and power supplies. In everyday life analogue systems are all around us; the medical thermometer, car speedometer and fuel gauge. The natural world is also essentially analogue, for example the transition from day into night in terms of ambient light (Fig. 5.2).

Analogue signals feature in communication systems such as the telephone where the output from a microphone might look like Fig. 5.3.

Therefore a signal, device, or system can be described as analogue if it varies through an infinite number of levels.

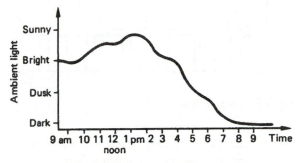

Fig. 5.2 Analogue transition from day to night

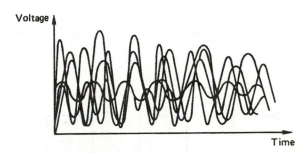

Fig. 5.3 Analogue microphone output

Digital signals

Such a signal is one that has finite levels. In Fig. 5.4 there is not an infinite number of levels but ten (0–9V). However, the most common digital signals have two levels: 'off' and 'on'. Typical applications of digital signals are the light switch and the digital watch. Digital signals are also used in communication systems, usually in the form of a code like

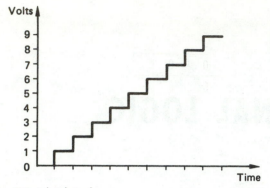

Fig. 5.4 A digital signal

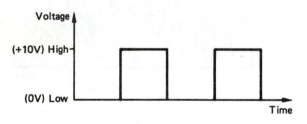

Fig. 5.5 Morse code signal

morse code (Fig. 5.5). Here it is the duration of the 'on' period that conveys the information:

short duration = 'dot',
long duration = 'dash'.

A particularly useful and very widely utilised digital signal is the *binary signal*, which has two levels (Fig. 5.6).

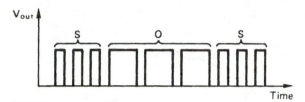

Fig. 5.6 Binary signal

LOGIC LEVELS

The low level is called Logic 0 and the high level Logic 1.

$$0V = Logic\ 0$$
$$+10V = Logic\ 1$$

Because the Logic 1 level is more positive than the 0 level, it is called *positive logic*. Similarly if the 0

level were $-5V$ and the 1 level 0V, this would also be positive logic. Conversely, if Logic 0 = $+10V$ and Logic 1 = 0V this would be *negative logic*. In this book only positive logic will be considered.

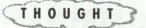

■ *So 0V = Logic 0 and +5V = Logic 1*
This is basically true, but in reality voltages are seldom exact, so in practice logic levels are represented by a band of voltage.

e.g.　　　　　　　　　$0–1V = Logic\ 0$
and　　　　　　　　　$4–5V = Logic\ 1$ *(Fig. 5.7).*

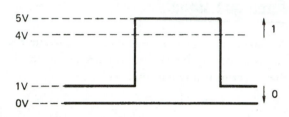

Fig. 5.7 Voltage bands

Logic gates

A logic gate is a circuit that may have a number of inputs but has only one output, which will be either Logic 1 or Logic 0 as determined by the inputs.

Example Consider the circuit shown in Fig. 5.8. If any switch, A or B or C is set to Logic 1 (i.e. closed) the lamp will light. This gate is called the OR gate, simply because if A *or* B *or* C are at Logic 1 the output will be at Logic 1.

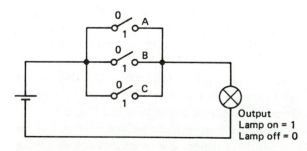

Fig. 5.8 OR gate circuit

Likewise in Fig. 5.9, the lamp will only light if all inputs are at Logic 1. Consequently this is called

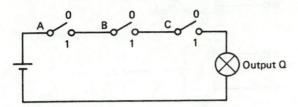

Fig. 5.9 AND gate circuit

an *AND gate* because the output will only be at Logic 1 when inputs A *and* B *and* C are at Logic 1.

THE TRUTH TABLE

This is one method of determining how the output of a gate will change according to the state of its inputs. Consequently a table must be drawn up that shows every permutation of input conditions – the *Table of Truth*.

Consider the circuit shown in Fig. 5.10, and its truth table.

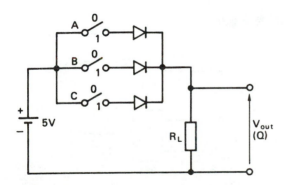

Fig. 5.10 Truth table circuit

If anything below 1.0V is Logic 0 and everything above 4.0V is Logic 1, from the table it can be seen that this circuit is an OR gate.

Note The way the table is drawn up, input C changes every line, B changes every two lines, A every four. For a three input gate there are eight possible input combinations (binary: $2^3 = 8$). For a four input gate there would be 16 input combinations (2^4). For a two input gate there would be only four combinations (2^2).

Table 5.1 Truth table for Fig. 5.10

A	B	C	Q (V)	Logic value
0	0	0	0	0
0	0	1	4.4	1
0	1	0	4.4	1
0	1	1	4.4	1
1	0	0	4.4	1
1	0	1	4.4	1
1	1	0	4.4	1
1	1	1	4.4	1

LOGIC GATE SYMBOLS AND TRUTH TABLES

Just as with all electronic components and circuits, a conventional form of representation exists for Logic gates. Unfortunately the British Standards Institute Symbol (BSI) is different from the international symbol (MIL, ANSI) and while the ANSI symbol is now accepted as the standard form of representation you will need to be aware of the BS symbol (Fig. 5.11).

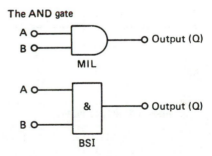

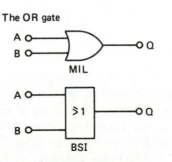

Fig. 5.11 Logic gate symbols

Table 5.2 AND gate truth table

A	B	Q
0	0	0
0	1	0
1	0	0
1	1	1

Table 5.3 OR gate truth table

A	B	Q
0	0	0
0	1	1
1	0	1
1	1	1

The NOT gate

This is a single input gate that gives an output that is the opposite of the input – it is an inverter.

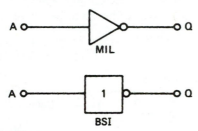

Fig. 5.12 NOT gate

Table 5.4 Not gate truth table

A	Q
0	1
1	0

It seems reasonable that if the output of any logic gate is fed into an inverter or NOT gate, the output of the NOT gate will be the inverse of the logic gate (Fig. 5.12 and Table 5.4), i.e. an AND gate followed by a NOT gate will be a NOT–AND gate, called a *NAND* gate. Likewise, an OR gate followed by a NOT gate will be a NOT–OR gate, called a *NOR* gate.

The NAND gate (Fig. 5.13)

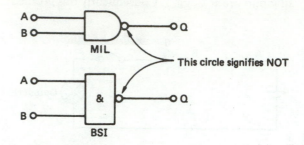

Fig. 5.13 NAND gate

Table 5.5 NAND gate truth table

A	B	Q	
0	0	1	
0	1	1	
1	0	1	
1	1	0	← NOT 'A and B'

Note This output is the exact opposite of the AND gate.

The NOR gate

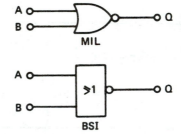

Fig. 5.14 NOR gate

Table 5.6 NOR gate truth table

A	B	Q	
0	0	1	
0	1	0	
1	0	0	
1	1	0	NOT A or B

This is the opposite of the OR gate.

The discrete circuit AND gate

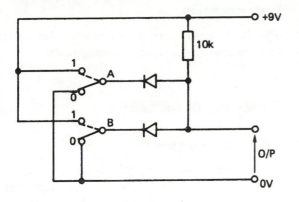

Equipment
Power supply
Breadboard
Voltmeter (digital or analogue)
2 general purpose Si diodes
10k resistor

Method
1 Build the circuit as shown.
2 Connect the power supply and adjust for 9V d.c.
3 Draw up a truth table for every possible input condition.

4 Complete the truth table by monitoring the output with the voltmeter and connecting inputs A and B to Logic 0 (0V) and Logic 1 (+9V) as appropriate.

The discrete circuit NOT gate

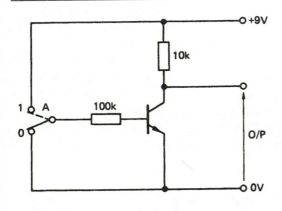

Equipment
Power supply
Breadboard
Voltmeter (digital or analogue)
General purpose Si n-p-n transistor, e.g. BC108
10k, 100k resistor

Method
1 Build the above circuit.
2 Connect the power supply and adjust for 9V d.c.
3 Connect the input A to Logic 0 (0V) and record the output voltage V_{out} with the voltmeter.

4 Connect the input A to Logic 1 (+9V) and record the output voltage.
5 Draw up and complete a truth table for the circuit.

Discrete and integrated logic circuits

You have built and investigated some discrete logic gates: a diode resistor AND gate and a resistor transistor NOT gate. These, whilst being quite simple, are relatively bulky. Because it is usual to require many gates when constructing logic circuits, it makes sense to use integrated circuit technology.

The first integrated circuits used resistor transistor logic (RTL) followed by diode transistor logic (DTL) and today these have been replaced by Transistor–Transistor Logic (TTL), Complementary Metal Oxide Semiconductor Logic (CMOS) and Emitter Coupled Transistor Logic (ECL).

So it is possible to buy an integrated circuit logic gate in TTL, CMOS and ECL.

WHAT IS THE DIFFERENCE?

To consider this question we need to be aware of the requirements placed on an IC logic gate. They can be specified as the following properties or parameters:

- fan-in
- fan-out
- propagation delay time
- noise margin
- power consumption

Fan-in

This is the number of gates that can be connected to the input of an IC gate without affecting its performance.

Fan-out

This is the maximum number of gates that can be connected to the output of an IC gate without causing its output to change from the specified 0 and 1 levels.

Propagation delay time

This is time taken for the output of a gate to change after an input has been applied.

Noise margin

Noise refers to unwanted or interference signal voltages. The noise margin is the maximum noise voltage that can be tolerated at the input of a gate without the output changing.

Power consumption

This is the amount of power taken by one gate under quiescent and operating conditions.

COMPARISON OF THE GATES

Table 5.7 Table of gate properties

Type of gate	Fan-in	Fan-out	Propagation delay (ns)	Noise margin (V)	Power consumption (mW)
TTL	8	10	9	0.4	40
CMOS	8	50	30	1.5	1
ECL	5	50	1.1	0.4	30

As you can see from Table 5.7, for speed of operation ECL is the fastest and this is why it finds application in systems like mainframe computers where speed is of the essence.

For commonplace uses, TTL and CMOS are utilised. With CMOS offering the advantage of good fan-out capability and low power consumption – particularly important where battery powered circuitry is involved because it gives improved battery life – the logic Practical Investigations included in this book use CMOS 'chips' but TTL equivalent ICs can be used if they are more readily available.

> ## THOUGHT
>
> - *So when building a circuit, provided the correct gates are used, a mixture of TTL and CMOS chips can be utilised?*
>   *No! Because of the difference between the two types of integrated circuit it is always best to use only one type; a mix of types could lead to unreliable or faulty operation.*

Logic gate selection

TTL

Standard TTL gates are available in the 74 series, i.e. 7400 is a Quad 2 input NAND gate. This is a single integrated circuit that contains four 2 input NAND gates (Fig. 5.15).

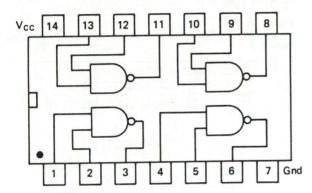

Fig. 5.15 TTL integrated circuit logic gate

In addition to the standard TTL gates, manufacturing processes enable the TTL gate to be made offering some of the refinements normally only available with CMOS technology

i.e. 74 F   series – fast switching speed
74 HC   series – high speed low power consumption
74 HCT series – high speed CMOS replacement for LSTTL
74 LS   series – low power improved speed
74 ALS series – advanced low power (twice the speed and half the power consumption of 74 LS).

All the available choices seem to be confusing, but remember the various alternatives merely offer improvements on the standard TTL gate. The connections and basic operation are the same. If you study the manufacturer's selection guide, it can be seen that the code number is the same, but the prefix letters are different and not all gates are available in all types.

For example, a Quad 2 input NAND gate is available in all types and coded; the 7400 in

standard TTL, or 74F00 as fast TTL or 74ALS00 as advanced low power TTL.

A dual four input AND gate 7421 is not available in standard TTL but only as a 74LS21 or 74HCT21.

Practical selection

1 First decide the type and number of gates you require in your circuit. i.e. 3 two-input OR gates, 2 inverters (NOT gates), 1 three-input NAND gate.

2 Look for the gate you require and note the code number. A two-input OR is available as a Quad 2 input IC (four gates) code 7432. NOT gate – is available as a Hex (six) package code 7404. 3 input NAND gate is available as a triple 3 input IC code 7410.

So to build your circuit requiring six gates you will need three integrated circuits 7432, 7404, 7410.

T H O U G H T

■ These ICs contain more gates than the circuit requires – does that matter?
No. You simply use what you require and isolate the unused gates (refer to the practical assignments).

Gate connections

The pin connections for the common 74 series ICs are given and must be used when connecting circuits using this series.

CMOS

CMOS logic gates are available in the 4000B series and while the catalogue selection may appear much smaller than that for the 74 series all the common gates are available. For example, the 4011B is a Quad 2 input NAND gate and like the 7400 contains four two-input NAND gates (Fig. 5.16).

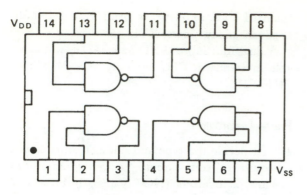

Fig. 5.16 The 4011B integrated circuit logic gate

CAUTION! At first glance the connections appear identical to the 7400 but the gates on the 7400 point one way, on the 4011B they point into the middle of the IC. Therefore connections for TTL and CMOS are different!

The common CMOS IC gate pin connections are given and must be used for the practical assignments.

Practical assignments using logic gates

CIRCUIT CONNECTION

Probably the best means of building IC circuits is to use a breadboard, which is a plug-in matrix board marketed under various trade names, allowing speedy accurate construction (Fig. 5.17).

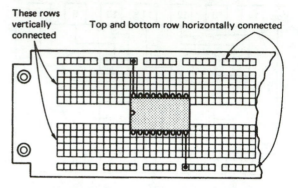

Fig. 5.17 A breadboard

PRACTICAL INVESTIGATION 22

The OR and AND gate

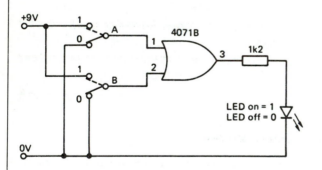

Equipment
Power supply (or 9V battery)
Breadboard
One output LED
1k2 resistor
Two input switches
4081 B (AND), 4071 B (OR)
Connecting wire

Method

1 Examine the diagram of the 4071 B and you will see it contains four OR gates. Connect it to make the circuit shown above.

2 Adjust the power supply to give 9V and connect it to the IC ensuring +9V goes to V_{DD} and 0V to V_{SS}.

3 *Note*   You are only using one gate; it is a good idea to connect the input pins of the other three (5, 6, 8, 9, 12, 13) to the $+V_e$ or 0V rail.

4 By investigation complete the truth table.

5 Switch off the supply and change the IC for a 4081 B (check the connections are correct).

6 Repeat the investigation and complete the truth table.

OR				AND		
A	B	Q		A	B	Q
0	0			0	0	
0	1			0	1	
1	0			1	0	
1	1			1	1	

INPUTS AND OUTPUTS

Inputs

A number of input switches will be required, each capable of supplying Logic 0 or Logic 1:

Logic 0 = supply 0V level,
Logic 1 = supply $+ V_e$ level.

Note A logic 0 must be connected to the 0V level and not left floating (Fig. 5.18).

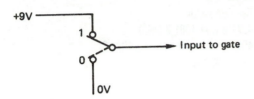

Fig. 5.18 Logic circuit input

In practice, individual switches can be used or dual in line (d.i.l.) switches are available in the same format as an integrated circuit, or a wire connection can simply be plugged into the $+ V_e$ or 0V line as required to give a Logic 1 or 0 respectively.

Output

The best output indicating device is a LED with a series resistor to limit the current.

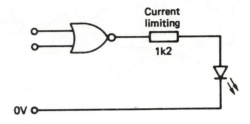

Fig. 5.19 Logic circuit output

As with the input switches, individual LEDs can be used or a 10 bar LED array can be obtained in a dual in line (d.i.l.) package like an integrated circuit.

PRACTICAL INVESTIGATION 23

The NOR and NAND gate

Equipment
Power supply
Breadboard
One output LED
1k2 resistor
Two input switches
4001 B (NOR), 4011 B (NAND)
Connecting wire

Method
1 Connect the circuit as shown using the 4001 B. (Ensure polarity is correct for the 9V power supply.)
2 By investigation complete the truth table shown opposite.
3 Replace the 4001 B with a 4011 B and complete the truth table.

Result
Compare these completed truth tables with those of Practical Investigation 22.

NOR				NAND		
A	B	Q		A	B	Q
0	0			0	0	
0	1			0	1	
1	0			1	0	
1	1			1	1	

Combined gates 1

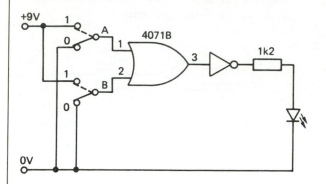

Equipment
Power supply
Breadboard
One output LED
1k2 resistor
Two input switches
4071 B (OR) 4081 B (AND)
4069 B (inverter alternative = 4049 B)
Connecting wire

Method
1 Build the circuit as shown above. If a 4049 B Hex inverter is used in place of the 4069 B the pin connections are different (see diagram).
2 Draw a truth table and complete it by investigation.
3 Replace the 4071 B with a 4081 B and complete a truth table.

Results
Compare the truth tables produced with those of Practical Investigation 23 and draw a conclusion about the circuits involved.

Combined gates 2

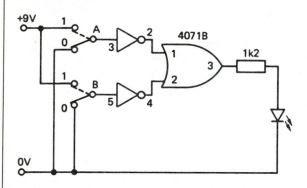

Equipment
Power supply
Breadboard
One output LED
1k2 resistor
Two input switches
4071 B, 4081 B
4069 B (4049 B)
Connecting wire

Method
1 Build the circuit as shown above using two gates of the Hex inverter.
2 Draw up a truth table and complete by investigation.
3 Replace the 4071 B with a 4081 B and repeat the investigation completing a truth table.

Results
Compare the resulting two truth tables with those from Practical Investigations 23 and 24 and draw a conclusion from your studies.

Power supply

CMOS integrated circuits offer the best noise immunity and highest quality operating speed with a supply of 9–13V.

A 9V battery or a suitable stabilised d.c. supply can be used.

Note Be sure that you always connect your IC to the power supply rails: $+9V$ to V_{DD}, 0V to V_{SS}.

CAUTION! Check first with the pin connection diagram; some 16 pin ICs have different power supply pins.

Unused gates

CMOS devices can be damaged by high voltage static. Because the human body can store a charge of many thousands of volts it makes sense to protect the unused gates by connecting the inputs to the positive or negative power supply rail.

OBSERVATIONS ON THE PRACTICAL INVESTIGATIONS

You will have discovered that your work with the AND, OR, NAND, NOR gates simply verifies the theoretical truth tables, but in addition to this you will have made the following important findings.

1 An OR gate followed by a NOT gate gives a NOR function.
2 An AND gate followed by a NOT gate gives a NAND function.
3 An OR gate preceded by NOT gates gives a NAND function.
4 An AND gate preceded by NOT gates gives a NOR function.

Therefore a NAND and NOR function can be obtained by inverting the output of an AND and OR gate or by inverting the inputs of an OR and AND gate.

SELF ASSESSMENT 9

1 Using truth tables decide the output of:
   a) a two input NAND gate followed by a NOT gate;
   b) a two input NOR gate followed by a NOT gate.
2 Could a NAND and a NOR gate be used as a NOT gate?

3 What is the significance of these answers?
4 Draw the truth table for a three-input AND gate.

Open collector gates

If you study the TTL data sheets at the back of this book you may notice that a number of gates are available with 'open collector outputs' – e.g. the 7401 is the same as the 7400 – it is a quadruple two-input NAND gate but it has open collector outputs. Likewise the 7409 is the same as the 7408 quadruple two-input AND gate but is of the open collector variety!

There are two main applications of this device.

1 To enable a TTL gate to provide an output other than 5.0V. You may remember that the TTL logic levels are nominally 0V and $+5.0V$. This is quite a problem when interfacing a TTL gate with a circuit or device that requires a larger than 5.0V input signal. The open collector gate can be used to overcome this difficulty.
2 To allow the outputs of TTL gates to be connected together, i.e. 'paralleled' in order to perform a different logic function.

Let us use a practical example to illustrate the use of an open circuit TTL gate. Suppose a TTL gate was required to supply an input signal to a CMOS gate that had a $+12.0V$ supply. The logic levels for the CMOS inputs will be approximately: *logic $0 = 0V$ logic $1 = +12V$*. The output from the TTL gate could not possibly operate this CMOS gate. An open collector gate can be used as shown in Fig. 5.20.

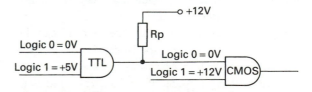

Fig. 5.20 Logic level change using open collector gates

The external 'pull-up' resistor (Rp) is connected to the CMOS supply rail of $+12V$. When the TTL output is Logic '0' the CMOS input is 0V. When the TTL output is Logic '1' the CMOS input will be +12V – *problem solved!*

- Can you tell from the logic symbol if a gate has an open collector output?
  Yes, a convention exists for this. Fig. 5.21 shows the representations used, a half circle or an asterisk indicates an open collector output. However these are by no means universal, the only true indication is given by the code number and the data sheet.

Fig. 5.21 Open collector representation

WARNING! Do not attempt to use normal TTL gates to drive CMOS circuits. Also, do not connect in parallel the outputs of conventional TTL or CMOS gates. They do not have open circuit collectors and the resulting excessive current flow will destroy the gates.

Tristate logic

Open collector logic requires the use of pull-up resistors, this tends to degrade the speed of operation and the noise immunity of the gates. A solution to this problem is to use 'Tristate' logic gates. This term is very misleading because it implies that three logic levels are used. This is not so! Tristate gates have the same function as conventional gates except that the output can be switched into a high impedance state using the control input. Fig. 5.22 shows tristate versions of the OR and AND gate. If the control input is at Logic '1' they behave normally. When the control input is taken to Logic '0' the gate outputs assume a high impedance state. This facility allows the tristate device to be connected along with others to a common line or 'bus'. Such devices are often used in data selection and highway-based systems.

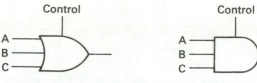

Fig. 5.22 Tristate logic gates

Assertion level logic notation

It is worth mentioning at this stage a peculiarity concerning logic levels.

POSITIVE LOGIC

The more positive voltage level = Logic '1'
The more negative voltage level = Logic '0'

e.g.   0V = '0'   +5V = '1'

NEGATIVE LOGIC

This is the reverse of positive logic thus:

the more negative voltage level = Logic '1'
the more positive voltage level = Logic '0'

e.g.   −5V = '1'   0V = '0'.

A gate or device that operates from a Logic '1' can be described as being 'active high' while a Logic '0' triggered device is said to be 'active low'. The standard logic symbol assumes positive or 'active high' logic. To indicate that a device or gate is 'active low' or logic '0' triggered it is conventional to use negating circles on the inputs to the device as shown in Fig. 5.23. This is assertion level logic.

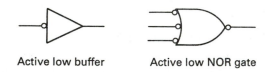

Active low buffer   Active low NOR gate

Fig. 5.23 Assertion level logic

Boolean notation

The way a logic gate behaves can be determined by a truth table, but while this is very convenient for gates with two inputs it becomes very unwieldy for more than two inputs simply because the number of lines follows the law of 2^n where n = number of inputs. Therefore, whilst a two-input gate requires 2^2 (4) lines, a four-input gate requires 2^4 (16) lines! Life also becomes complicated when analysing a circuit into a number of interconnected gates. To simplify matters, *Boolean notation* is used which gives a Boolean expression for each type of gate. This is named after the nineteenth century mathematician George Boole who did a great deal of pioneering work on logic and computers long before the advent of electronics.

The OR gate

The name of the gate gives an indication of its behaviour; if the inputs A or B are at Logic 1, a Logic 1 is obtained at the output. In Boolean algebra OR is represented by a plus sign (+). Therefore A or B is written A + B (Fig. 5.24).

Fig. 5.24 Boolean notation for the OR gate

The output expression Q = A + B tells us exactly how the gate performs and the output conditions required for Logic 1 output.

The AND gate

The output of a two-input AND gate will be Logic 1 if both inputs are at 1 (if A AND B are at 1). Boolean notation for AND is a dot. Therefore A AND B is written A · B (Fig. 5.25).

Fig. 5.25 Boolean notation for the AND gate

The NOT gate

You are aware that the function of a NOT gate is to invert. Therefore the output is always the opposite of the input. Boolean notation for inversion is a bar over the top of the expression.

So, $\overline{A}$ indicates that the output is NOT A (Fig. 5.26).

Fig. 5.26 Boolean notation for the NOT gate

The NAND gate

The name tells us we will *not* get an output when inputs A AND B are at Logic 1. Because this is the inverse of the AND gate it is represented by a bar over the AND expression (Fig. 5.27).

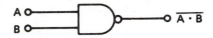

Fig. 5.27 Boolean notation for the NAND gate

The NOR gate

This is the inverse of the OR gate and so can be represented by a bar over the OR expression (Fig. 5.28).

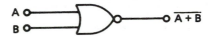

Fig. 5.28 Boolean notation for the NOR gate

A word about Boolean expressions

We are dealing with a topic called logic but it may seem to readers that things are becoming far from logical! This is because the mind is easily confused by different uses for common symbols, namely the signs used in Boolean algebra.

A + B means A OR B but we recognise the plus sign (+) and it is easy to read A + B as A and B. Similarly, A · B means A AND B so although the dot (·) represents multiplication in mathematics, in logic it must be read as AND.

This is of course mathematically illogical but the notation is that used by Boole originally and today it is the normal notation for logic circuits.

USE OF BOOLEAN EXPRESSIONS

Consider circuits 1 and 2 and 3 overleaf.

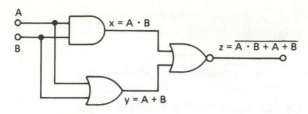

Fig. 5.29 Boolean circuit 1

Table 5.8 Truth table for Boolean circuit 1

A	B	x	y	z
0	0	0	0	1
0	1	0	1	0
1	0	0	1	0
1	1	1	1	0

$A \cdot B$ $\quad$ $A + B$

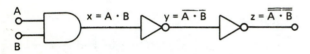

Fig. 5.30 Boolean circuit 2

Table 5.9 Truth table for Boolean circuit 2

A	B	x	y	z
0	0	0	1	0
0	1	0	1	0
1	0	0	1	0
1	1	1	0	1

$A \cdot B$ $\quad$ $\overline{A \cdot B}$ $\quad$ $\overline{\overline{A \cdot B}} = A \cdot B$

It is obvious that a double inversion sign cancels out.

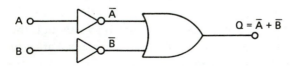

Fig. 5.31 Boolean circuit 3

THOUGHT

■ *Does this mean that* $\overline{A} + \overline{B} = \overline{A \cdot B}$?
Yes it does.

This is an example of a very important and useful concept in logic systems known as DeMorgan's Laws.

DEMORGAN'S LAWS

The first

This states that $\overline{A \cdot B} = \overline{A} + \overline{B}$.

Table 5.10 Truth table for Boolean circuit 3

A	B	A	B	Q
0	0	1	1	1
0	1	1	0	1
1	0	0	1	1
1	1	0	0	0

From your previous investigations you will recognise the output Q as a NAND function.

This is easily proved by Table 5.11.

Table 5.11 Truth table for DeMorgan's first law

A	B	$\overline{A}$	$\overline{B}$	$\overline{A} + \overline{B}$	$A \cdot B$	$\overline{A \cdot B}$
0	0	1	1	1	0	1
0	1	1	0	1	0	1
1	0	0	1	1	0	1
1	1	0	0	0	1	0

QED

The second

This states that $\overline{A + B} = \overline{A} \cdot \overline{B}$, which is easily proved by Table 5.12.

Table 5.12 Truth table for DeMorgan's second law

A	B	$\overline{A}$	$\overline{B}$	$\overline{A} \cdot \overline{B}$	$A + B$	$\overline{A + B}$
0	0	1	1	1	0	1
0	1	1	0	0	1	0
1	0	0	1	0	1	0
1	1	0	0	0	1	0

QED

Now use Practical Investigations 26 and 27 to prove these laws for yourself.

3 Write the Boolean expression for the following circuit:

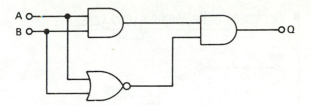

SELF ASSESSMENT 10

1 Write the Boolean expression for the following circuit:

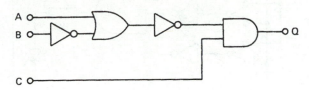

2 Draw the truth table for the following circuit:

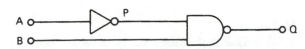

4 If $Q = \overline{A + B} + C$, write an equivalent expression using DeMorgan's Laws.

Combinational logic networks

From the work encountered in the previous section you will have realised that it is possible to interconnect logic gates in order to provide any output function. This facility enables logic circuits

PRACTICAL INVESTIGATION 26

Verification of DeMorgan's Laws (1)

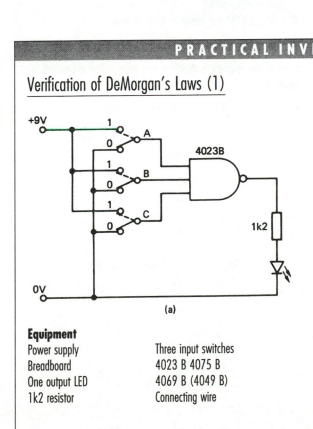

(a)

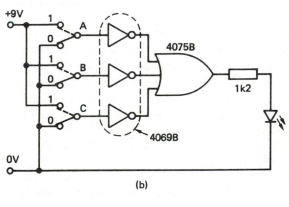

(b)

Equipment

Power supply	Three input switches
Breadboard	4023 B 4075 B
One output LED	4069 B (4049 B)
1k2 resistor	Connecting wire

Method

1 Build the circuit (a) shown above, draw up and complete the truth table.

2 Build circuit (b) and complete the truth table by investigation.

Results

Compare the two truth tables and write a Boolean expression linking them.

PRACTICAL INVESTIGATION 27

Verification of DeMorgan's Laws (2)

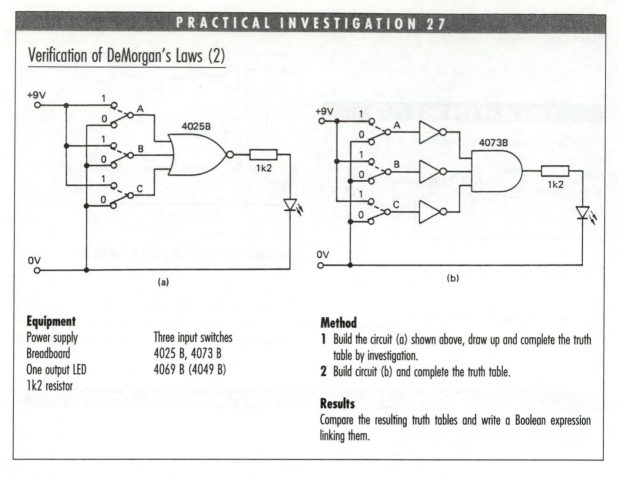

(a)   (b)

Equipment

Power supply   Three input switches
Breadboard   4025 B, 4073 B
One output LED   4069 B (4049 B)
1k2 resistor

Method

1 Build the circuit (a) shown above, draw up and complete the truth table by investigation.
2 Build circuit (b) and complete the truth table.

Results

Compare the resulting truth tables and write a Boolean expression linking them.

---

to be used for a great number of applications including machine interlocks, burglar alarms, vending machines, combination locks etc.

To design a combinational logic circuit it is necessary to discover the input and output requirements of a particular application. To illustrate this, consider the following examples.

Example 1

In order for a piece of machinery to be operated safely the following conditions must be satisfied.

1 The workpiece is in position (microswitch closed giving a Logic 1).
2 The safety guard is correctly positioned (sensor operates giving a Logic 1).
3 The operator is in the correct position (infra-red beam broken giving Logic 1).

Only under these conditions will power be available at the starter of the machine; under all other conditions the starter will not operate.

A truth table (Table 5.13) can be drawn to represent the system where

A = workpiece switch,
B = safety guard sensor,
C = infra-red beam sensor,
Q = power to the starter.

Table 5.13 Truth table for Example 1

A	B	C	Q
0	0	0	0
0	0	1	0
0	1	0	0
0	1	1	0
1	0	0	0
1	0	1	0
1	1	0	0
1	1	1	1

Only under the single condition $Q = A \cdot B \cdot C$ will the starter operate the machine. Consequently, this circuit could be fashioned using an AND gate. This fact can be discovered by drawing up a truth table for the problem and then deriving a Boolean expression for the condition that satisfies all the requirements.

Example 2

Access to a compound that contains dangerous high voltage equipment can be gained by a maintenance electrician under the following conditions.

1 The high voltage is off (Logic 0).
2 A keyswitch on the control panel 100 yds away is off (Logic 0).
3 A keyswitch on the gate is turned on (Logic 1).

Under all other conditions the gate cannot physically be opened.

Interpret the logic requirements for this arrangement and draw up a truth table and derive a Boolean expression.

A = high voltage
B = control panel switch
C = gate switch

Conditions for entry: $Q = 1$
Requirements are $A = 0$, $B = 0$, $C = 1$.
The truth table is shown in Table 5.14.

Table 5.14 Truth table for Example 2

A	B	C	Q
0	0	0	0
0	0	1	1
0	1	0	0
0	1	1	0
1	0	0	0
1	0	1	0
1	1	0	0
1	1	1	0

$$\therefore Q = \overline{A} \cdot \overline{B} \cdot C.$$

Now design a combinational logic network that meets this requirement.

This is a comparatively easy task because for this circuit there is only one possible output condition (Fig. 5.32). Where there are a number of permitted outputs the situation is somewhat different.

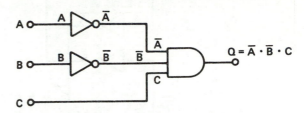

Fig. 5.32 Combinational logic circuit for Example 2

Consider Table 5.15.

Table 5.15 Truth table with four permitted outputs

A	B	C	Q	
0	0	0	0	
0	0	1	1	$(\overline{A} \cdot \overline{B} \cdot C)$
0	1	0	1	$(\overline{A} \cdot B \cdot \overline{C})$
0	1	1	0	
1	0	0	0	
1	0	1	1	$(A \cdot \overline{B} \cdot C)$
1	1	0	1	$(A \cdot B \cdot \overline{C})$
1	1	1	0	

By examination of the table you can see that there are four permitted output conditions and the overall Boolean expression is

$$Q = \overline{A} \cdot \overline{B} \cdot C + \overline{A} \cdot B \cdot \overline{C} + A \cdot \overline{B} \cdot C + A \cdot B \cdot \overline{C}$$

Incidentally, it is okay to leave out the dot ($\cdot$) when writing a Boolean expression; it then becomes

$$Q = \overline{A}\,\overline{B}\,C + \overline{A}\,B\,\overline{C} + A\,\overline{B}\,C + A\,B\,\overline{C}$$

However it is written, it is clear that the expression is long and contains a number of interconnected gates. In addition to a truth table, a Boolean expression can be obtained from a logic circuit diagram directly by writing the inputs and outputs to each gate to build up the complete Boolean expression, as shown in Fig. 5.33.

The Boolean expression for this combinational logic circuit is given as

$$Q = \overline{\overline{A \cdot \overline{B}} \cdot \overline{\overline{A} \cdot B}}$$

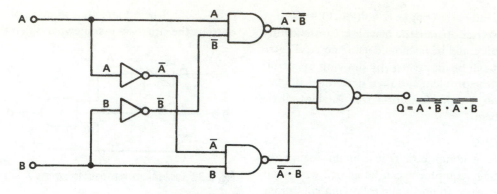

Fig. 5.33 A combinational logic circuit showing how the Boolean output expression is derived

Whichever way a Boolean expression is derived it is often very complicated and not always in its simplest form. For this reason it is wise to simplify or minimize an expression before dealing with the situation any further.

Minimisation techniques

The methods used to simplify Boolean expression are:

1 Boolean algebra,
2 Karnaugh maps.

BOOLEAN ALGEBRA

Algebraic simplification is helped by the use of Logic identities. These are simply sets of rules that specify a logic gate's output for a given input condition.

AND gate

$A \cdot 0 = 0$ For all values of A, because the other input is 0, output = 0.

$A \cdot 1 = A$ If $A = 0$, the output will be 0. If $A = 1$, output = 1.

$A \cdot A = A$ If $A = 1$, output = 1. If $A = 0$, output = 0.

$A \cdot \overline{A} = 0$ If $A = 1$, $\overline{A} = 0$. Therefore output will be 0.

OR gate

$A + 0 = A$ If $A = 1$, inputs = $1 + 0$ so output = A.

$A + 1 = 1$ If $A = 1$, inputs = $1 + 1$ so output = 1. If $A = 0$, inputs = $0 + 1$. Therefore output = 1.

$A + A = A$ If $A = 1$, inputs = $1 + 1$ so output = A.

$A + \overline{A} = 1$ If $A = 0$, $\overline{A} = 1$ inputs = $0 + 1$ so output = 1.

Combination of gates

OR

$$(A + B) + C = A + (B + C)$$
$$(A \cdot B) + (A \cdot C) = A \cdot (B + C)$$

AND

$$(A \cdot B) \cdot C = A \cdot (B \cdot C)$$
$$(A + B) \cdot (A + C) = A + (B \cdot C)$$

DeMorgan's Laws

These two laws have already been investigated and proven to be extremely powerful tools for minimizing logic functions:

$$\overline{A \cdot B} = \overline{A} + \overline{B}$$
$$\overline{A + B} = \overline{A} \cdot \overline{B}$$

To apply the rule invert each variable and change the logic sign:

e.g. $$\overline{P + Q} = \overline{P} \cdot \overline{Q}$$

Do not forget that a double inversion cancels:

$$\therefore \overline{\overline{P+Q}} = P + Q$$

So now let us look again at the expression

$$Q = \overline{\overline{A \cdot \overline{B} \cdot \overline{A} \cdot B}}$$

Apply DeMorgan's Laws to remove the top inversion bar and change the sign:

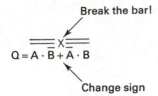

Break the bar!

$$Q = A \cdot \overline{B} + \overline{A} \cdot B$$

Change sign

We now have a double inversion which cancels giving:

Cancel

$$Q = A \cdot \overline{B} + \overline{A} \cdot B$$

So the simplified expression is

$$Q = A \cdot \overline{B} + \overline{A} \cdot B$$

What about the other long expression

$$Q = \overline{A} \cdot \overline{B} \cdot C + \overline{A} \cdot B \cdot \overline{C} + A \cdot \overline{B} \cdot C + A \cdot B \cdot \overline{C}?$$

Examine this for common factors:

$$\overline{A}\,\overline{B}C + \overline{A}B\overline{C} + A\overline{B}C + AB\overline{C}$$

$\overline{B} \cdot C$ common $B \cdot \overline{C}$ common

$$Q = \overline{B}\,C(\overline{A} + A) + B\,\overline{C}(\overline{A} + A)$$
since $(\overline{A} + A) = 1$
$$\therefore Q = \overline{B}C + B\overline{C}$$

Mapping techniques

These methods of minimisation are very quick and convenient to use and with practice considerable skill can be acquired. Perhaps the most popular method is the Karnaugh map (pronounced *car-no*).

For a two-variable expression four squares are required:

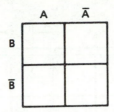

This is the map and a Boolean expression can be entered on to it by entering a Logic 1 in the appropriate boxes.

Take the example $Q = A \cdot B + A \cdot \overline{B}$:

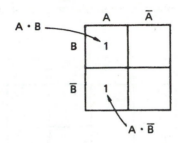

The entire Boolean expression has now been mapped and any 'common' cells can be grouped.

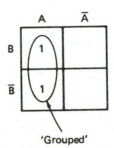

'Grouped'

The only term common to these two cells is A. So, the expression has been simplified to $Q = A$.

From the truth table shown below devise a Boolean expression and then simplify it using Karnaugh mapping techniques.

A	B	Q
0	0	0
0	1	1
1	0	1
1	1	1

For a situation involving three variables (A, B, C) the map would look like this

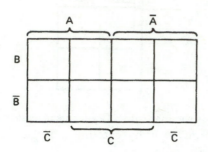

Each term is present in four squares or cells. So a map like this

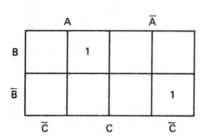

Would represent the Boolean expression

$$Q + A \cdot B \cdot C + \overline{A} \cdot \overline{B} \cdot \overline{C}$$

(it could not be further simplified).

Example
Simplify the following expression using a Karnaugh map.

$$Q = \overline{A} \cdot \overline{B} \cdot C + \overline{A} \cdot B \cdot \overline{C} + A \cdot \overline{B} \cdot C + A \cdot B \cdot \overline{C}$$

1 Draw up the map.

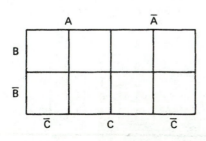

2 Insert the four terms according to the identity of the variables.

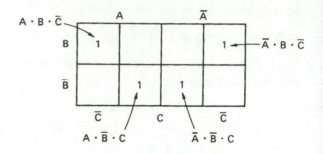

3 Group any common cells.

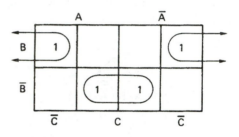

4 Determine the common variables to the linked cells. Two groups gives two terms.

$$\therefore Q = B \cdot \overline{C} + \overline{B} \cdot C$$

■ *So any cells can be grouped provided they are neighbours? Not quite, there are rules.*

Coupling rules

1 The number of cells must be a power of 2, i.e. 1, 2, 4, 8 etc.
2 For the simplest expression the maximum number must be grouped.
3 A cell can appear in more than one group.
4 Cells must have a common edge. The map can be imagined as a sphere opened out (just like the map of the world). This gives the following coupling possibilities and many more:

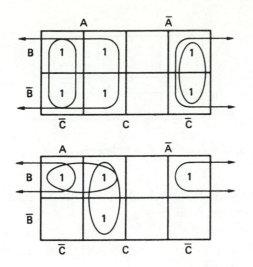

SELF ASSESSMENT 12

1 A logic circuit requirement has the truth table shown below. Derive the Boolean expression and then simplify using a Karnaugh map.

A	B	C	Q
0	0	0	0
0	0	1	1
0	1	0	1
0	1	1	1
1	0	0	0
1	0	1	0
1	1	0	1
1	1	1	0

2 Minimise the following Boolean expression using a Karnaugh map.

$$Q = A \cdot B \cdot \overline{C} + A \cdot B \cdot C + A \cdot \overline{B} \cdot \overline{C} + A \cdot \overline{B} \cdot C$$

Logic circuit design

Provided a Boolean expression can be obtained, it is a relatively simple matter to simplify it as required, and then interconnect the appropriate gates to provide the desired output. For example, $Q = \overline{A} \cdot C + B \cdot \overline{C}$ can be represented by the circuit shown in Fig. 5.34.

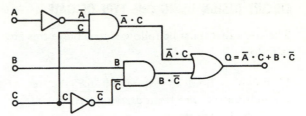

Fig. 5.34 Circuit representation of $Q = \overline{A} \cdot C + B \cdot \overline{C}$

To implement this practically will require

- 2 inverters,
- 2 AND gates,
- 1 OR gate.

Whilst this is perfectly feasible it is not always practical to mix gates because of availability. It is far more convenient to build a circuit from one type of gate only – either NANDs or NORs. This can be achieved using DeMorgan's Laws and common sense. For example, $Q = \overline{A} \cdot B$ implemented using NAND gates only is shown in Fig. 5.35.

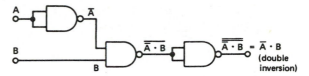

Fig. 5.35 Circuit representation of $Q = \overline{A} \cdot B$ using NAND gates

$Q = \overline{A} \cdot B$ using only NOR gates is shown in Fig. 5.36.

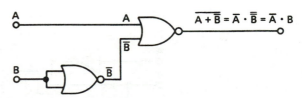

Fig. 5.36 Circuit representation of $Q = \overline{A} \cdot B$ using NOR gates

THOUGHT

- This is hardly minimisation! Does the application of DeMorgan's Law lead to greater complexity?
  In some cases it does when it is required to implement a circuit using only one gate type, but since integrated circuits contain a number of gates it is unimportant.

CIRCUIT DESIGN USING ONE TYPE OF GATE

The steps that must be followed to achieve this are as follows:

1 Obtain a Boolean expression for the set of conditions, either from drawing a truth table or simple investigation.

Example
Three switches A, B, C control a device. The device must operate if A is on and B is on but C is off *or* if A is off, B is on and C is on *or* if A is on, B is off and C is on. Under all other conditions the device must not operate.

A	B	C	Q	
0	0	0	0	
0	0	1	0	
0	1	0	0	
0	1	1	1	A off B on C on
1	0	0	0	
1	0	1	1	A on B off C on
1	1	0	1	A on B on C off
1	1	1	0	

The Boolean expression is

$$Q = \overline{A} \cdot B \cdot C + A \cdot \overline{B} \cdot C + A \cdot B \cdot \overline{C}$$

2 Simplify the Boolean expression if possible using the Karnaugh map shown in Fig. 5.37.

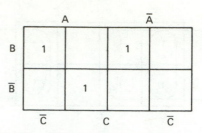

Fig. 5.37 Karnaugh map of the Boolean expression

$Q = \overline{A} \cdot B \cdot C + A \cdot \overline{B} \cdot C + A \cdot B \cdot \overline{C}$ is in its simplest form.

·3 Design the circuit using AND, OR and NOT gates (Fig. 5.38).
4 To implement using NAND gates only replace each gate with its NAND equivalent (Fig. 5.39).
5 Examine the circuit for redundant gates, e.g. two NANDS in series, then re-draw (Fig. 5.40).

A SPECIAL LOGIC GATE

Consider the truth table below giving the Boolean expression $Q = \overline{A} \cdot B + A \cdot \overline{B}$.

A	B	Q
0	0	0
0	1	1
1	0	1
1	1	0

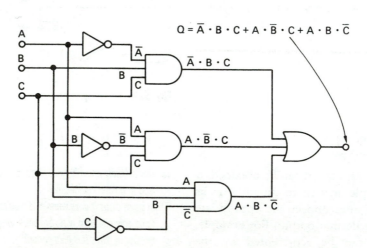

Fig. 5.38 Circuit design using AND, OR and NOT gates

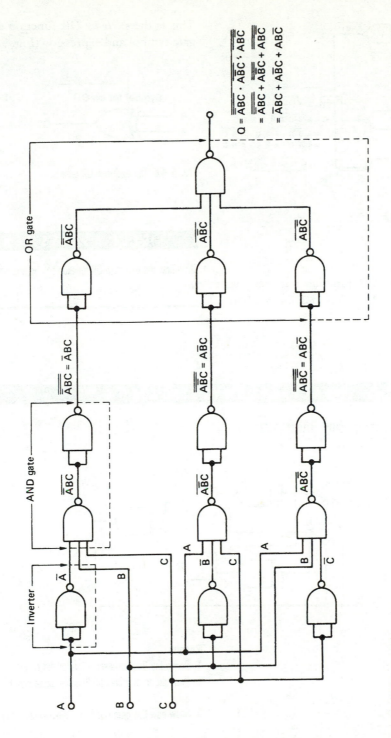

Fig. 5.39 Circuit design using NAND gates only

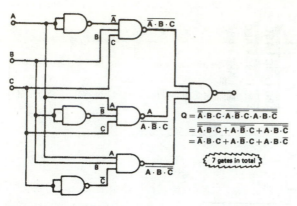

$$Q = \overline{\overline{A \cdot B \cdot C} \cdot \overline{A \cdot \bar{B} \cdot C} \cdot \overline{A \cdot B \cdot \bar{C}}}$$
$$= \overline{\overline{A \cdot B \cdot C} + \overline{A \cdot \bar{B} \cdot C} + \overline{A \cdot B \cdot \bar{C}}}$$
$$= \bar{A} \cdot B \cdot C + A \cdot \bar{B} \cdot C + A \cdot B \cdot \bar{C}$$

7 gates in total

Fig. 5.40 Final circuit design

This is similar to the OR function but there will only be a 1 output if A or B is at 1 but *not* both.

This is the *exclusive* OR function and has its own gate symbol and expression (Fig. 5.41).

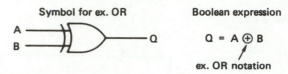

Fig. 5.41 The exclusive OR gate

Show how the exclusive OR function can be implemented using only NOR gates.

The exclusive OR function using NAND gates

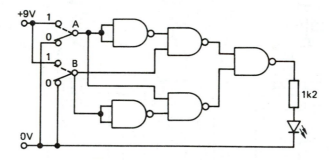

Equipment
Power supply (or 9V battery)
Breadboard
Output LED
1k2 resistor
Two input switches
Two 4011B NAND gates

Method
1 Interconnect the gates to produce the circuit shown.
2 By investigation produce a truth table for the circuit.

Results
1 Derive the Boolean expression for the circuit.
2 Draw the truth table and Boolean expression for an exclusive NOR gate.
3 Show how this gate could be implemented using only NAND gates.

Combinational logic review

- An analogue signal is one that has an infinite number of possible levels.
- A digital signal is one that has a finite (specific) number of levels.
- A binary signal is a digital signal that has only two levels; high and low or Logic '1' and Logic '0'.
- A logic signal where the high or more positive voltage level represents Logic '1' is *positive logic.*
- A logic signal where the more negative voltage represents Logic '1' is *negative logic.*
- A logic gate is a circuit that may have many inputs but only one output, e.g. Logic 1 or 0 as determined by the inputs.
- Logic gates are available in integrated circuit form as' TTL, ECL or CMOS packages.
- There are circuit symbols for all the logic gates. Today the international symbol (MIL, ANSI) is universally used, but the BS symbol is still to be found in many publications and circuits in this country.
- TTL gates are available with 'Open collector' outputs that allow:
  a) gate outputs to be wired together to perform a specific logic function;
  b) a standard TTL logic output signal (+5V) to be connected to a device or circuit operating from a higher voltage.

- A truth table can be used to determine the behaviour of any logic gate or system.
- Boolean notation is used to simplify logic circuits without the use of truth tables.
- DeMorgan's Laws show how a logic circuit can be fabricated using gates of only one type.
- Assertion level logic can be used to show when a device operates or is switched using the more negative logic level (active low).
- A problem or situation can be interpreted using Boolean expressions.
- A Boolean expression can be derived from a truth table or logic circuit diagram.
- Long unwieldy Boolean expressions can be simplified using;
  a) Boolean algebra,
  b) mapping techniques.
- A Karnaugh map is a quick, convenient way of minimising a complex Boolean expression.
- Any Boolean expression can be implemented using NOR gates only or NAND gates only.
- An 'exclusive OR' gate is a special type of OR gate that proves valuable in many logic applications.

SELF ASSESSMENT ANSWERS

Self Assessment 9

1 a) NAND followed by a NOT = AND
   b) NOR followed by a NOT = OR

2 A NAND and NOR gate can be used as a NOT by joining all inputs together to make one (see below):

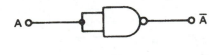

3 This means that any logic gate can be constructed using NANDs or NORs to create equivalent gate circuits, i.e. a circuit with 3 OR gates, 2 AND gates, 1 NOT gate and 5 NAND gates could be made entirely from NAND or NOR gates.

4

A	B	C	Q
0	0	0	0
0	0	1	0
0	1	0	0
0	1	1	0
1	0	0	0
1	0	1	0
1	1	0	0
1	1	1	1

Self Assessment 10

1 $Q = \overline{A + \overline{B} \cdot C}$

2

A	B	P	Q
0	0	1	1
0	1	1	0
1	0	0	1
1	1	0	1

3 $Q = A \cdot B \cdot \overline{A + B}$

4 $Q = \overline{A + B + C}$

   $\therefore Q = \overline{A} \cdot \overline{B} \cdot \overline{C}$

Self Assessment 11

Boolean expression = $Q = \overline{A} \cdot B + A \cdot \overline{B} + A \cdot B$

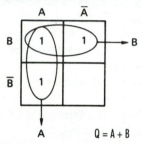

$Q = A + B$

Self Assessment 12

1 $Q = \overline{A}\overline{B}\overline{C} + \overline{A}B\overline{C} + \overline{A}BC + AB\overline{C}$

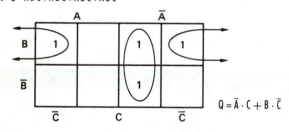

$Q = \overline{A} \cdot C + B \cdot \overline{C}$

2

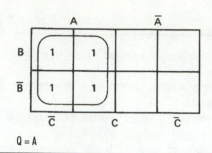

$Q = A$

Self Assessment 13

Exclusive OR function using NOR gates

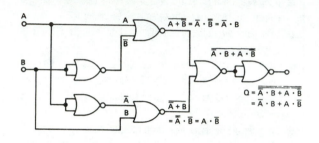

Multiple choice questions

1 The Boolean expression for the output of the circuit shown in Fig. 5.42 is given by:

   a) $Q = A \cdot B \cdot C + D$

   b) $Q = A + B + C \cdot D$

   c) $Q = A \cdot B + C + D$

   d) $Q = A \cdot B \cdot C \cdot D$

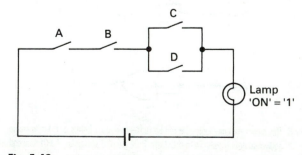

Fig. 5.42

2 The correct Boolean expression for the truth table shown in Fig. 5.43 is:
a) $Q = \overline{A} \cdot B \cdot \overline{C} + A \cdot \overline{B} \cdot C + A \cdot B \cdot C$
b) $Q = \overline{A} \cdot \overline{B} \cdot C + A \cdot \overline{B} \cdot C + A \cdot B \cdot \overline{C}$
c) $Q = \overline{A} \cdot \overline{B} \cdot C + \overline{A} \cdot B \cdot \overline{C} + A \cdot \overline{B} \cdot C$
d) $Q = \overline{A} \cdot \overline{B} \cdot \overline{C} + \overline{A} \cdot B \cdot \overline{C} + A \cdot \overline{B} \cdot C$

A	B	C	Q
0	0	0	0
0	0	1	1
0	1	0	1
0	1	1	0
1	0	0	0
1	0	1	1
1	1	0	0
1	1	1	0

Fig. 5.43

3 The Karnaugh map shown in Fig. 5.44 gives the following minimised expression:
a) $Q = A \cdot \overline{B} + \overline{C}$
b) $Q = B + \overline{A} \cdot \overline{C}$
c) $Q = B + C$
d) $Q = B + \overline{A \cdot C}$

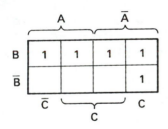

Fig. 5.44

4 The Boolean expression $Q = \overline{\overline{A \cdot B} + \overline{\overline{C} \cdot D}}$ can be modified using DeMorgan's Laws to give the equivalent expression:
a) $Q = A \cdot B \cdot \overline{C} \cdot \overline{D}$
b) $Q = A \cdot B + \overline{C \cdot D}$
c) $Q = A \cdot B \cdot \overline{C} + \overline{D}$
d) $Q = A + B \cdot \overline{C} \cdot \overline{D}$

5 The Boolean expression for the output of the logic circuit shown in Fig. 5.45 is given by:
a) $Q = \overline{A} + (B + A \cdot B \cdot C)$
b) $Q = A \cdot B \cdot C + (B \cdot \overline{A})$
c) $Q = A \cdot B \cdot C + (B + \overline{A})$
d) $Q = \overline{A} \cdot (A \cdot B \cdot C + B)$

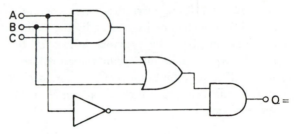

Fig. 5.45

6

SEQUENTIAL LOGIC

—

In a combinational logic network the state of the output (Q) is determined by the state of the input at that precise instant in time.

A sequential logic circuit is one where the output (Q) is determined not only by the state of the inputs but also the previous values of the input. A sequential logic circuit therefore has a memory. The basic building block of a sequential circuit is the *bistable*, which is a simple 1 bit memory device.

The bistable

This is a circuit that has two stable states. It will remain in a given state until it is triggered, when it will change state only to remain in this new state until triggered again. A bistable has two inputs and two outputs (Fig. 6.1).

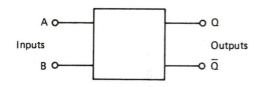

Fig. 6.1 The bistable

The outputs are self-explanatory really: whatever Q is $\overline{Q}$ is the inverse. Hence if Q = 1, $\overline{Q}$ = 0 or if Q = 0, $\overline{Q}$ = 1. Practical Investigation 29 illustrates this.

THE RS BISTABLE

The inputs to this bistable have the terms Set (S) and Reset (R). The logic symbol is shown in Fig. 6.2.

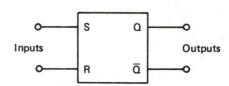

Fig. 6.2 Logic symbol of RS bistable

When Q = 1 ($\overline{Q}$ = 0) the device is said to be *Set*. When Q = 0 ($\overline{Q}$ = 1) the device is said to be *Reset*.

Now, because this circuit is a simple memory device, the output will be determined by (a) the input conditions, and (b) what the output was before the new input conditions were applied.

To set the circuit (to make Q = 1) a logic pulse must be applied to S, and R must be Logic 0, but if the circuit was already set before the pulse was applied on S, the circuit will remain in the Set state. To draw the truth table the existing state of the output is called Q and the state after the application of a Set or Reset pulse is called Q + 1. Hence Table 6.1 can be drawn.

Table 6.1 RS bistable truth table

S	R	Q	Q+1	
0	0	0	0	No pulse on S or R; Q stays as it was.
0	0	1	1	No pulse on S or R; Q stays as it was.
1	0	0	1	Pulse on S sets Q to 1.
1	0	1	1	Since Q = 1 pulse on S = no change.
0	1	0	0	Since Q = 0, pulse on R = no change.
0	1	1	0	Pulse on R resets Q to 0.
1	1	0	X	Indeterminate
1	1	1	X	Indeterminate

Note The indeterminate state means that when S = R = 1, the output Q could be 1 or 0. Therefore the situation where S = R = 1 is not allowed, i.e. if both inputs are at 1, there is an equal probability that the output will be 1 or 0. An RS flip-flop can be constructed from NOR and NAND gates as Practical Investigations 30 and 31 will show, but be warned – they operate in slightly different ways!

'Clocking' a logic circuit

It is often very convenient for the input conditions to a flip-flop to change, but the output not to change until signal or clock pulse has been applied to the circuit. In this way all flip-flops in a sequential circuit can, if required, operate simultaneously. If an RS flip-flop were to be provided with a clock input it might look like Fig. 6.3.

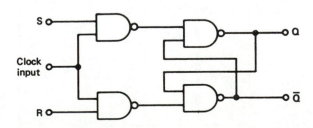

Fig. 6.3 RS flip-flop with clock input

Regardless of the RS input conditions the output will not change until a clock pulse has been applied (Table 6.2).

Table 6.2 Truth table for an RS flip-flop with a clock input

S	R	Clock	Q	Q+	
0	0	0	0	0	No clock pulse (clock inactive) = no change
0	0	0	1	1	
0	1	0	0	0	
0	1	0	1	1	
1	0	0	0	0	
1	0	0	1	1	
0	0	1	0	0	Clock pulse but no change since S = R = 0
0	0	1	1	1	
0	1	1	0	0	Reset S = 0 R = 1
0	1	1	1	0	
1	0	1	0	1	Set S = 1 R = 0
1	0	1	1	1	

The remaining combinations will produce an indeterminate result.

The JK flip-flop

The indeterminate states of the RS flip-flop make it unsuitable for many applications and for this reason the JK flip-flop has emerged as being very popular. To understand the main difference it is necessary to compare its truth table with that of the RS flip-flop (Tables 6.3 and 6.4).

Table 6.3 RS flip-flop truth table

S	R	Q	Q+
0	0	0	0
0	0	1	1
1	0	0	1
1	0	1	1
0	1	0	0
0	1	1	0
1	1	0	x
1	1	1	x

The transistor bistable

Equipment
Power supply
Breadboard
Two LEDs
Two 470R resistors
Three 10k resistors
Two BC 108 transistors

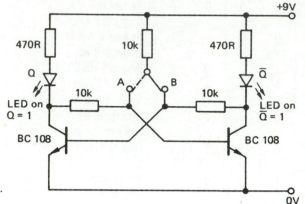

Method
1 Connect up the circuit as shown.
2 Switch on the supply and by investigation complete the truth table.

A	B	Q	$\overline{Q}$
Pulse	–		
–	Pulse		

The 2 NOR gate RS flip-flop

Equipment
Power supply
Breadboard
4001
Two output LEDs
Two 1k2 resistors
Two input switches

Method
1 Connect the circuit as shown using two of the Quad 2 input NOR gates.
2 By investigation, complete the truth table.

S	R	Q	$\overline{Q}$
0	1		
0	0		
1	0		
0	0		
0	1		
0	0		
1	1		

The 4 NAND gate RS flip-flop

Equipment

Power supply	Two output LEDs
Breadboard	Two 1k2 resistors
4011 B	Two input switches

Method

1 Build the above circuit using the four gates in a Quad 2 input NAND gate.
2 By investigation complete the following truth table.

S	R	Q	$\bar{Q}$
0	1		
0	0		
1	0		
0	0		
0	1		
0	0		
1	1		

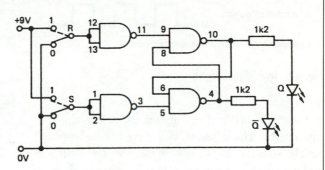

Results

Compare the resultant truth table with that of the NOR gate flip-flop and hence explain the difference between the two.

Table 6.4 JK flip-flop truth table

J	L	Q	Q+
0	0	0	0
0	0	1	1
1	0	0	1
1	0	1	1
0	1	0	0
0	1	1	0
1	1	0	1 change
1	1	1	0 change

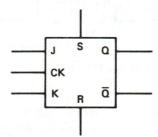

Fig. 6.4 BS symbol for JK flip-flop

Remember the Q+ condition indicates the state of the Q output *after* a J or K pulse has been applied to the input. You can see that the output Q of the flip-flop always changes state when both the J and the K inputs are applied to the device. This means that this flip-flop has a *toggle mode*, which means that if both inputs J and K are held at Logic 1 then the output Q will change its state after every clock pulse.

Note The facility exists for setting and resetting the flip-flop as well as the clock input.

Examine the pin connection diagram for the dual JK flip-flop integrated circuit No. 4027B and familiarise yourself with its layout before attempting the investigation into the JK flip-flop.

A word about sequential circuits

The majority of such circuits have to be 'clocked' or 'strobed'. To do this a logic signal must be applied to the clock input line of the circuit (Fig. 6.5).

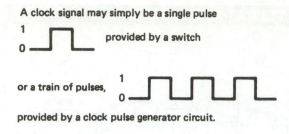

A clock signal may simply be a single pulse

1
0   provided by a switch

or a train of pulses, **1** **0**

provided by a clock pulse generator circuit.

Fig. 6.5 Types of clock signals

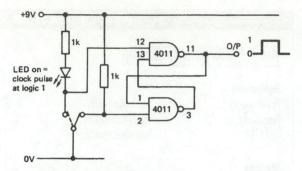

Fig. 6.7 The debounced switch

Generally when investigating the behaviour of a circuit the clock pulse will have to be applied singly, so a switch is the best method. *But* switches suffer from a phenomenon known as *contact bounce*: this means that although the switch may be switched on and then off to provide a single pulse the actual voltage wave form looks like Fig. 6.6.

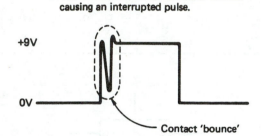

The mechanical switch contacts 'bounce' causing an interrupted pulse.

+9V

0V

Contact 'bounce'

Fig. 6.6 Illustration of contact bounce

The circuit may interpret this waveform as a number of pulses! Consequently it is a good idea to use a *contact bounce suppressor circuit*.

THE DEBOUNCED SWITCH

For all sequential circuit investigations that require clocking, build the circuit shown in Fig. 6.7 at one end of the breadboard so that you have a guaranteed clean, noiseless, bounce-free pulse.

From your deliberations over the clocked JK flip-flop the following facts will have emerged. In conjunction with the clock:

1 At any stage if R is taken to Logic 1, Q resets to 0 and $\overline{Q}$ to 1, regardless of the previous state.
2 At any stage if S is taken to Logic 1, Q sets to 1 and $\overline{Q}$ to 0, regardless of the previous state.

3 It can if required remain in one particular state despite clocking. Therefore it has a memory.
4 If J = K = 1 the output will 'toggle', i.e. two clock pulses are required to change the output and then change it back. Therefore it can divide by two.

This 'divide by two' characteristic is a major advantage and can best be shown by referring to a waveform diagram as in Fig. 6.8. By using this facility flip-flops can be connected to form binary counters and shift registers. Practical Investigations 33 and 34 will allow you to prove this for yourself.

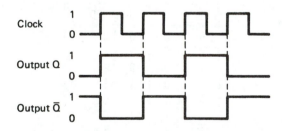

Fig. 6.8 Waveform diagram of a clocked JK flip-flop in 'toggle' mode

It can now be seen that these simple bistables can be interconnected and used in complex counter and shift register circuits.

Further sequential logic circuits

A glance at the data sheets included at the end of this book will show you that it is not always necessary to connect bistables together in order to provide the circuit you require, since they are often available ready made. For example, the TTL 7493

PRACTICAL INVESTIGATION 32

The JK flip-flop

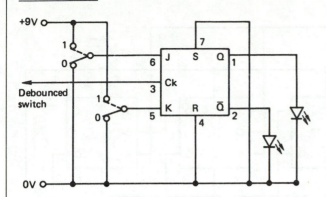

Equipment
Power supply
Breadboard
Two LEDs
4027 (JK flip-flop)
Connecting leads
Debounced switch
Two switches

Method

1 Build the circuit as shown ensuring that both S and R are connected to the 0V rail (S = R = 0).

2 By investigation complete the truth table below.

J	K	Clock	Q	$\bar{Q}$
0	0	⎍		
0	1	⎍		
1	0	⎍		
1	1	⎍		

For each condition, pulse the clock input a number of times and observe the outputs.

3 When you are familiar with the operation of the circuit, investigate the result of setting S to 1 and R to 1 for different JK input conditions.

Results

Make a general statement about the operation of the JK flip-flop indicating possible uses for it and the advantage of the Set, Reset facility.

---

is a '4 bit binary counter' while the CMOS 40161 is also a '4 bit counter'.

These devices are purpose built 'synchronous' counters that operate slightly differently to the asynchronous or 'ripple through' types previously encountered. With an asynchronous circuit it takes time for the pulse to 'ripple through' as each flip-flop toggles. This limits the speed at which the circuit can operate: if the circuit is operated at or near its maximum frequency there is a risk that an error will occur (this is often called a race hazard or glitch!). With a synchronous counter the clock pulses are applied to *all* stages simultaneously, thus improving the speed and reliability.

Have a look at the selection of counters that are shown below and then carry out investigation 35.

TTL synchronous counters

74161: 4 bit counter with direct clear

74162: 4 bit binary coded decimal (BCD) decade counter

74163: 4 bit binary counter with synchronous clear

74169: 4 bit up/down counter

74193: 4 bit up/down with dual clock

74168: BCD up/down counter

CMOS synchronous counters

4029: up/down binary/BCD counter

40161: 4 bit counter

40163: 4 bit counter

40193: 4 bit up/down with separate clocks

40160: BCD counter

40192: BCD up/down counter with separate clocks

PRACTICAL INVESTIGATION 33

The binary counter

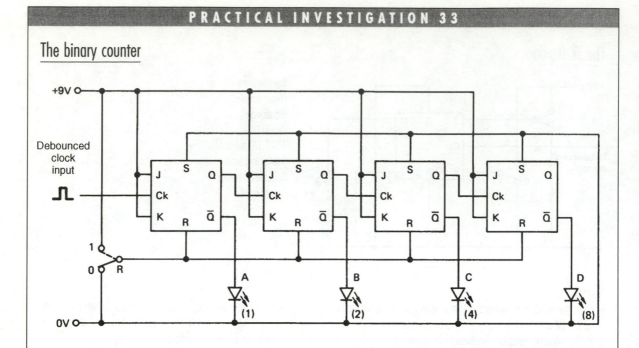

Equipment
Power supply
Breadboard
Two 4027s (JK flip-flops)
Four LEDs
Debounced switch
One switch
Connecting wire

Method
1 Build the circuit as shown making sure that all the set inputs are connected to the 0V line (Logic 0).
2 Set the Reset switch to 0.
3 Pulse the clock input repeatedly and observe the LEDs as each pulse is applied.
4 When a number of LEDs are ON switch the Reset to 1 and back to 0 and note the effect.

5 When you are familiar with the operation complete a table that will show the way counting is achieved.
NB Start count with all LEDs off.

	LED A (1)	LED B (2)	LED C (4)	LED D (8)
	0	0	0	0
Pulse 1	1	0	0	
Pulse 2	0	1		
Pulse 3	1	1		

Results
From your completed table deduce the maximum denary number that this circuit will count up to.

Providing clock pulses

So far clock pulses have been provided by manually pressing a button for every pulse that is required. It is usual for digital circuits to be switched using a regular train of pulses. Normally these pulses occur at very high frequencies, e.g. 1MHz (1 million pulses per second!). For our purposes a much lower frequency is required.

PRACTICAL INVESTIGATION 34

The shift register

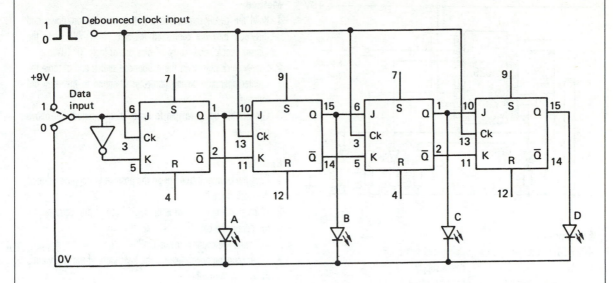

Equipment

Power supply
Breadboard
Two 4027s (JK flip-flops)
One 4049 (Hex inverter)

Four LEDs
Debounced switch
One switch
Connecting wire

Method

1 Build the circuit as shown above.
2 Connect all S and R inputs to the 0 V line.

3 Using the data input switch enter 4 bits into the register.
4 Operate the debounced clock pulse switch once and observe what happens to the stored information.
5 Operate the clock pulse switch a further three times and note carefully what happens to the LED output indicators.
6 Repeat the above for different sets of data.

Results

1 Determine the difference between a counter and a shift register.
2 Suggest an application where a shift register could be used.

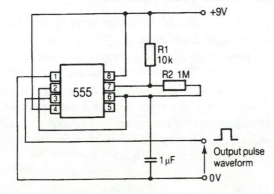

Fig. 6.9 Clock pulse generator

The circuit shown in Fig. 6.9 uses a 555 integrated circuit as a pulse generator, this produces pulses at a frequency of approximately 1 Hz (1 pulse per second). This creates a very useful and cheap pulse generator: you might consider building it on a separate breadboard and using it for operating counter and shift register circuits.

SHIFT REGISTERS

Just as counters are available in integrated circuit (IC) form so are shift registers, for example:

PRACTICAL INVESTIGATION 35

The synchronous binary counter

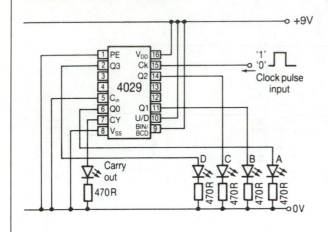

Equipment

4029 CMOS counter   Debounced switch
Five LEDs   Power supply
Five 470R resistors   Logic breadboard

Method

1 Build the circuit shown above noting that the up/down and BIN/BCD pins are connected to logic '1' (+9V) while the 'preset enable' and 'carry in' pins are at logic '0' (0V).

2 Supply clock pulses via the debounced switch and observe the counter operation paying particular attention to the carry out terminal.

3 Connect the up/down pin to logic '0' and repeat the above procedure.

Results

1 Does the counter increment on the positive or negative edge of the clock pulse?

2 When counting up, what is the logic state of the 'carry out':
   a) during the count?
   b) when the count is maximum?

3 When counting down, what is the logic state of the 'carry out':
   a) during the count?

TTL shift registers

7495: 4 bit serial/parallel input, parallel output; shift left/right

negative edge triggered

7496: 5 bit serial/parallel right, parallel output

negative edge triggered

74198: 8 bit, universal shift register

positive edge triggered

CMOS shift registers

4014: 8 bit parallel input, serial output

positive edge triggered

4015: dual 4 bit, serial input, parallel output

positive edge triggered

4035: 4 bit parallel input, parallel output

positive edge triggered

Notice that the triggering is specified as 'positive' or 'negative'. This refers to the edge of the

clock pulse that actually causes the triggering; thus, a positive edge triggered device is operated by the positive going edge of the trigger pulse.

Practical Investigation 36 allows you to check the operation of a shift register.

Seven segment displays

By now you should be quite familiar with the standard light emitting diode (LED) since this has been used to monitor the output of most of the logic circuits. The physical construction of this device is different from a normal diode. The light is emitted from the semiconductor junction so the maximum junction area must be exposed. For the standard LED a ring format is used for the junction as shown in Fig. 6.10a.

PRACTICAL INVESTIGATION 36

Parallel input/parallel output shift register

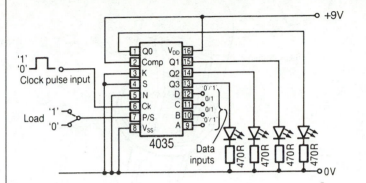

Equipment
4035 CMOS shift register
Four LEDs
Four 470R resistors
Debounced switch
Power supply
Logic breadboard

Method

1 Build the circuit shown above.
2 Notice that there are four parallel data inputs (A, B, C, D) and four outputs (Q0, Q1, Q2, Q3). Enter the data word 1100 into the register by connecting the appropriate input to Logic '1' (+9V) and Logic '0' (0V), e.g. A = +9V, B = +9V, C = 0V, D = 0V.
3 With the load switch set to Logic '0' operate the clock pulse several times, observe that *nothing happens!*
4 Set the load switch to Logic '1' and operate the clock pulse switch several times.
5 Now set the load switch to Logic '0' and operate the clock pulse switch. Hopefully you will have discovered that data are loaded into the register by a Logic '1' on the load input and *one* clock pulse. Further clock pulses do not change the output. When the load is set to Logic '0', data in the register are shifted by each clock pulse.

6 Change the input data word by making the appropriate connections and investigate the shifting operation. If you have a low frequency pulse generator use this to clock the register.
7 During the clocking sequence note the effect of setting the 'R' input to Logic '1' and back to Logic '0'.
8 Repeat the investigation with the 'comp' (pin 2) input connected to Logic '0'.

Results

1 What is the purpose of the 'R' input?
2 What is the purpose of the 'comp' input?

The area of the diode that emits the light is very small (about the size of a pin head). To improve the contrast and give a better viewing angle a coloured plastic lens is used. This causes the emitted rays of light to diverge and creates the impression of a larger light source. The construction of the LED is shown in Fig. 6.10b.

With digital systems it is often desirable to display decimal numbers, e.g. a binary coded decimal (BCD) counter counts from 0–9 in binary, but a decimal display is required rather than a binary display. The plastic lens system used on LEDs determines the actual shape that the eye sees, so it is possible to use seven LEDs and arrange them in the pattern shown in Fig. 6.11.

Each segment is allocated a specific letter and, by energising the appropriate segments, any number from 0 to 9 can be displayed. Thus, if segments a, b, c, d and g are energised the number 3 will be shown.

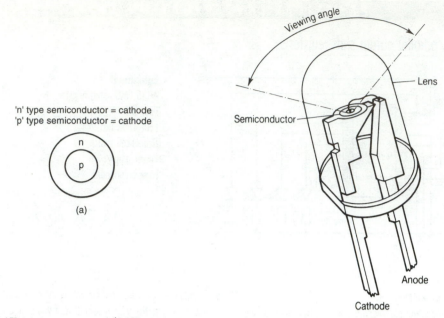

'n' type semiconductor = cathode
'p' type semiconductor = cathode

n

p

(a)

Viewing angle

Lens

Semiconductor

Anode

Cathode

(b)

Fig. 6.10 (a) LED junction arrangement. (b) LED construction

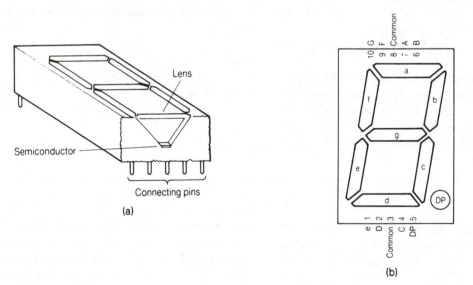

Lens

Semiconductor

Connecting pins

(a)

a

f   b

g

e   c

d   DP

(b)

Fig. 6.11 (a) Seven segment display construction. (b) Seven segment pattern

Common anode and common cathode displays

Seven segment displays contain seven LEDs in each display. Each of these diodes has an anode and a cathode making 14 connections in total. By taking a common point (either the anode or cathode) the number of connections can be reduced to eight as shown in Fig. 6.12a and b.

Notice that the polarity required to operate the device is different. The choice of display will be determined by how it is to be connected in the

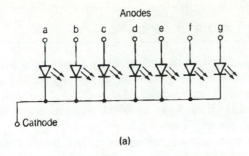

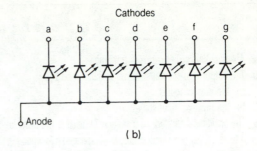

Fig. 6.12 (a) Seven segment common cathode. (b) Seven segment common anode

circuit and the type of logic used, e.g. positive or negative logic.

Practical Investigation 37 allows you to investigate the seven segment display by using a decoder integrated circuit to drive the display.

This converts the binary coded decimal signal into one that energises the correct segments of the display, e.g. 0101 is a binary code for denary number 5. The decoder energises segments a, c, d, f, g when the input is 0101.

It now becomes possible, by linking together available ICs and display devices and/or relays, to provide logic circuits that provide specific functions, for example:

a traffic light controller
a lift call system
a combination lock
electronic dice

You might care to devise your own circuits for these and many other applications, but if not there are many books and magazines that are devoted to designing and building digital systems such as these.

PRACTICAL INVESTIGATION 37

BCD using a seven segment display

Equipment
4511 CMOS BCD to seven segment decoder/driver
Standard seven segment display (common cathode)
Eight 470R resistors
Power supply
Logic breadboard

Method

1 Build the circuit shown above (check that your seven segment display pin connections are the same as those shown in the diagram).
2 Draw up a truth table showing how the binary input relates to the displayed number (input 'A' is the least significant bit – LSB).
3 Check that the circuit does provide a display for numbers 0 to 9.
4 By investigation discover what happens when you set the inputs to numbers greater than 9.

Sequential logic review

- A sequential circuit possesses a memory.
- The output of a sequential circuit is determined by:
  a) the present inputs,
  b) the previous input values.
- The basic building block of a sequential circuit is the bistable.
- A bistable has two stable states; it will change state when triggered and remain in the new state until triggered again.
- A bistable is a '1' bit memory device.
- A bistable or flip-flop has two inputs and two outputs.
- The outputs of a bistable are the complement of each other.
- The RS flip-flop has an indeterminate state when S = R = '1' – this is the *forbidden input condition*.
- The JK flip-flop is similar to the RS but will 'toggle' when J = K = '1'.
- The bistable is a 'divide by two' counter when in 'toggle' mode.
- The JK flip-flop is equipped with Set, Reset and clock inputs.
- Flip-flops can be interconnected to form counters and shift registers.
- It is often convenient to clock or pulse a sequential circuit. Whatever the input condition the output will not change until a clock pulse is received.
- A clock pulse can be produced by a simple on–off switch, provided it is debounced to prevent mechanical contact chatter.

- Sequential circuits are usually operated by a regular train of pulses produced by a pulse generator. The 555 timer integrated circuit can be used to provide such clock pulses.
- Counter and shift registers are available as purpose built integrated circuits in both TTL and CMOS form.
- Asynchronous counters are described as 'ripple through' devices since each bistable triggers the next and the pulse travels through the circuit rather like a line of dominoes with each one tipping the next in line.
- Asynchronous counters have a limited speed of operation.
- Synchronous counters apply the clock pulses to all stages simultaneously, this increases the speed of operation and the reliability.
- Counters and registers are triggered by the *edge* of the clock pulse, consequently it is possible to obtain positive and negative edge triggered devices.
- Light emitting diodes (LEDs) are available in seven segment form that allow denary (decimal) numbers 0 to 9 to be displayed.
- Seven segment displays require a decoder/driver stage to convert the binary signals into a form that gives a denary display.
- The vast selection of digital integrated circuits that are available allows specific or customised digital systems to be easily designed and built.

Multiple choice questions

1 A RS flip-flop has S = 0, R = 0 and the output $\overline{Q}$ = 1. If a pulse is now applied to S what will be the Q output?
 a) 0
 b) 1
 c) remain the same
 d) indeterminate

2 A RS flip-flop has S = 0, R = 1. A pulse is now applied so that S = 1 and R = 1. What will be the state of the $\overline{Q}$ output?
 a) 0
 b) 1
 c) remain the same
 d) indeterminate

3 A JK flip-flop has J = 1, K = 1 and Q = 1. What will be the $\overline{Q}$ output after three clock pulses have been applied?

 a) 1
 b) 0
 c) the same
 d) impossible to predict

4 A JK bistable is connected to operate in 'toggle' mode. If clock pulses are applied at a frequency of 100 Hz what will be the frequency of pulses at the Q output?
 a) 100 Hz
 b) 200 Hz
 c) 50 Hz
 d) 10 Hz

5 A seven segment display has segments a, b, c energised. What binary code does this represent?
 a) 1000
 b) 1010
 c) 0101
 d) 0111

7

DATA SHEETS

Small signal diodes

SIGNAL DIODES

Axial lead
a) L = 4.25, Dia = 1.85
b) L = 5.2, Dia = 2.7
d) L = 3.81, Dia = 1.71
e) L = 7.6, Dia = 2.7
f) L = 2.6, Dia = 1.7
Coloured band indicates cathode

TO-18

1	2	3
g)	k — a	
h)	a case k	
j)	a — k	

(k) TO-92

Application Code:
○ General Purpose — ◇ Switching — ● High speed
▽ VHF Tuner — □ Low leakage/low capacitance

Package	V_{RRM} max (V)	I_F (AV) max (mA)	V_F max (V)	@ I_F (mA)	App'n Code	Order Code
(e)	100	140	0.8	250	○○	AAZ15■
(e)	75	140	0.8	250		AAZ17■
(a)	30	100	1.1	100		BA317
(f)	35	100	1.2	100		BA482
(f)	125	225	1.0	200		BAS45
(a)	100	100	0.45	1	○○○	BAT41◆
(a)	30	100	0.4	10		BAT42◆
(a)	30	100	0.45	15		BAT43◆
(a)	100	150	0.45	10		BAT46◆
(a)	20	350	0.4	10		BAT47◆
(a)	40	350	0.4	10	○	BAT48◆
(b)	80	1000	0.42	100		BAT49◆
(f)	40	30	0.41	1	●	BAT81◆
(f)	60	30	0.41	1	●	BAT83◆
(f)	30	200	0.4	10	●	BAT85◆
(a)	60	300	1.0	200	●	BAV10
(a)	120	250	1.2	200	◇	BAV19
(a)	200	250	1.2	200	◇◇	BAV20
(a)	250	250	1.2	200	◇◇	BAV21
(g)	35	50	1.0	10		BAV45
(a)	75	100	1.0	100	●	BAW62
(a)	50	75	1.0	20	●	BAX13 01
(a)	150	200	1.3	100	○	BAX16
(a)	150	200	1.3	100	○○○○	BAX16 01
(a)	150	200	1.3	100	○○	BAX16ES
(a)	100	200	1.0	100	○	BAY72
(a)	100	200	0.8	10	○○○○	BAY73
(b)	20	1000	0.55	1000	○○○○○	BYV10-20◆
(b)	30	1000	0.55	1000	○○○	BYV10-30◆
(b)	40	1000	0.55	1000	○	BYV10-40◆
(b)	60	1000	0.7	1000		BYV10-60◆
(k)	35	10	1.5	5		JPAD100
(k)	35	10	1.5	5		JPAD50
(j)	35	50	1.5	5	□	PAD100
(h)	45	50	1.5	5	□	PAD5
(e)	30	110	0.54	130	○	OA47■
(e)	30	10	2.0	30	●	OA90 05■
(e)	115	50	2.1	30	○○	OA91 05■
(e)	115	50	1.85	30	○	OA95 05■
(a)	150	80	1.15	30		OA202 01
(e)	400V_{RW}	400	1.0	400	○	ZS104
(a)	100	75	1.0	10	●	1N914
(a)	100	75	1.0	10	●	1N916
(a)	125	200	0.8	10		1N3595
(a)	75	150	1.0	10	●	1N4148
(a)	75	200	1.0	10	●	1N4149
(a)	75	75	1.0	10	●	1N4150
(a)	50	200	0.74	10	●	1N4446
(a)	75	200	1.0	20	●	1N4448
(a)	75	150	1.0	100	●	1N4448-NSC
(a)	75	200	1.0	100	●	1S44
(d)	40	75	1.0	10	◇	1S920
(d)	50	200	1.2	200	◇◇	1S921
(d)	100	200	1.2	200	◇	1S922
(d)	150	200	1.2	200	◇	1S923
(d)	200	200	1.2	200	○	1S923

■ Germanium ◆ Silicon schottky barrier

Power diodes

Package (Not relative size)	IF(AV) Max Mean F'ward Current	50—100	200	300	400	600	800	1000	1200	1600
L=4.6, D=3.8 GLASS	1A				1N5060					
L=5, D=2.7 PLASTIC	1A	1N4001 (50V) / 1N4002 (100V)	1N4003		1N4004	1N4005	1N4006	1N4007		
		1N4001TR■ (50V) / 1N4002TR■ (100V)	1N4003TR■		1N4004TR■	1N4005TR■	1N4006TR■	1N4007TR■		
		1N4001GP◆ (50V) / 1N4002GP◆ (100V)	1N4003GP◆		1N4004GP◆	1N4005GP◆	1N4006GP◆	1N4007GP◆		
L=6.35, D=6.35 GLASS	3A		1N5624			1N5626				
L=8.9, D=3.7 PLASTIC	3A	30S1 (100V)	30S2		30S4	30S6	30S8	30S10		
L=9.65, D=5.3 PLASTIC	3A	1N5401 (100V)	MR502 / 1N5402		MR504 / 1N5404	1N5406		1N5408		
L=9.1, D=9.1 PLASTIC	6A	GI750 (50V) / GI751 (100V)	GI752			GI756				
L=9.5, D=6.35 PLASTIC	6A	60S1 DIODE (100V)	60S2		60S4	60S6	60S8	60S10		
TO-220	6.5A			BY249-300		BY249-600				

Selector chart of stud / lead mounting rectifier diodes. Each current rating has two lines: the upper line is the stud-cathode (**) version, the lower line the reverse-polarity (R) version. Part numbers are grouped by voltage (suffix = voltage rating).

Current	Thread	100V	200V	300V	400V	600V	800V	1000V	1200V	1600V
10A	10-32 UNF 2A			BYX98-300**		BYX98-600**			BYX98-1200**	
				BYX98-300R**		BYX98-600R**			BYX98-1200R**	
12A			12F20*					12F100*		
			12FR20*							
15A				BYX99-300**					BYX99-1200**	
16A		M16-100**	M16-200**		M16-400**		M16-800**		M16-1200**	
		M16-100R*	M16-200R*		M16-400R*		M16-800R*		M16-1200R*	
			16F20*				16F80**	16F100**	16F120**	
			16FR20*				16FR80*	16FR100*	16FR120*	
30A	METRIC M5			BYX96-300**		BYX96-600**			BYX96-1200**	BYX96-1600**
				BYX96-300R**		BYX96-600R**			BYX96-1200R**	BYX96-1600R**
40A (A)		M41-100**	M41-200**			M41-600**				
40A (A)		M41-100R*	M41-200R*			M41-600R*				
40A (A)		40HF10**	40HF20**		40HF40**	40HF60**	40HF80**	40HF100**	40HF120**	
40A (A)		40HFR10*	40HFR20*		40HFR40*	40HFR60*	40HFR80*	40HFR100*	40HFR120*	
47A (B)						BYX97-600**			BYX97-1200**	BYX97-1600**
47A (B)						BYX97-600R*			BYX97-1200R*	BYX97-1600R*
70A (A)		M71-100**	M71-200**		M71-400**	M71-600**	M71-800**			
70A (A)		M71-100R*	M71-200R*		M71-400R*	M71-600R*	M71-800R*			
70A (A)		70HF10**	70HF20**		70HF40**	70HF60**	70HF80**	70HF100**	70HF120**	
70A (A)		70HFR10*	70HFR20*		70HFR40*	70HFR60*	70HFR80*	70HFR100*	70HFR120*	
150A (A)		45L10**	45L20**		45L40**	45L60**	45L80**	45L100**	45L120**	
150A (A)		45LR10*	45LR20*		45LR40*	45LR60*	45LR80*		45LR120*	
250A (B)		70U10**	70U20**		70U40**				70U120**	
250A (B)		70UR10*	70UR20*						70UR120*	

Thread / mounting details:

- 70A: (A) 1/4-28 UNF 2A (B) METRIC M6
- 150A / 250A: (A) 1/2-20 UNF 2A (B) 5/8-16 UNF 2A

** Denotes Stud Cathode
* Denotes stud anode
■ Denotes Bandoliered
◆ Glass passivated hermetically sealed construction, with proven reliability equal to MIL-S-19500

Important — Forward current ratings quoted on stud mounting devices are maximum rating. Manufacturer's data should always be consulted as in some cases devices have to be forced air cooled to obtain the maximum ratings quoted.

Zener diodes

L = 4.5, D = 2.0   BZX55 Series — glass DO35
L = 4.25, D = 1.85   BZY79 Series — glass DO35

L = 12.5, D = 6.5
BZY70 Series — plastic
Rounded end indicates cathode

L = 4.8, D = 2.6   BZV85 Series — glass DO41
L = 5.2, D = 2.7   BZX85 Series — glass DO41

Axial lead types: Coloured band indicates cathode

L = 8.9, D = 3.7
1N5000 Series — plastic

BZY93 Series — 10/32 UNF 2A stud

L = 4.57, D = 3.81 BZT03 Series — glass SOD-57
L = 5.0, D = 4.5 BZW03 Series — glass SOD-64

BZY91 Series ¼ × 28 UNF Stud

\# These types are available as normal (stud cathode). Add suffix 'R' to Order Code if reverse polarity is required.

Mfr	Philips	SGS-Thomson	SGS-Thomson	Philips	Philips	Philips	—	Philips	Philips	Philips
Nominal Zener Voltage	WATTAGE (All +5% Voltage Tolerance)									
	400mW	500mW	1.3W	1.3W	2.5W	3W	5W	6W	20W	75W
2.4V	BZX79C2V4	BZX55C2V4								
2.7V	BZX79C2V7	BZX55C2V7	BZX85C2V7							
3V	BZX79C3V0	BZX55C3V0	BZX85C3V0							
3.3V	BZX79C3V3	BZX55C3V3	BZX85C3V3				1N5333B			
3.6V	BZX79C3V6	BZX55C3V6	BZX85C3V6				1N5334B			
3.9V	BZX79C3V9	BZX55C3V9	BZX85C3V9				1N5335B			
4.3V	BZX79C4V3	BZX55C4V3	BZX85C4V3				1N5336B			
4.7V	BZX79C4V7	BZX55C4V7	BZX85C4V7				1N5337B			
5.1V	BZX79C5V1	BZX55C5V1	BZX85C5V1	BZX85C5V1			1N5338B			
5.6V	BZX79C5V6	BZX55C5V6	BZX85C5V6	BZX85C5V6			1N5339B			
6.2V	BZX79C6V2	BZX55C6V2	BZX85C6V2	BZX85C6V2			1N5341B			
6.8V	BZX79C6V8	BZX55C6V8	BZX85C6V8	BZX85C6V8			1N5342B			
7.5V	BZX79C7V5	BZX55C7V5	BZX85C7V5	BZX85C7V5	BZX70C7V5	BZT03C7V5	1N5343B		BZY93C7V5#	BZY91C7V5
8.2V	BZX79C8V2	BZX55C8V2	BZX85C8V2	BZX85C8V2	BZX70C8V2	BZT03C8V2	1N5344B		BZY93C8V2#	
9.1V	BZX79C9V1	BZX55C9V1	BZX85C9V1	BZX85C9V1	BZX70C9V1	BZT03C9V1	1N5346B		BZY93C9V1#	
10V	BZX79C10	BZX55C10	BZX85C10	BZX85C10	BZX70C10	BZT03C10	1N5347B	BZW03-C10	BZY93C10#	BZY91C10
11V	BZX79C11	BZX55C11	BZX85C11	BZX85C11	BZX70C11	BZT03C11	1N5348B		BZY93C11	
12V	BZX79C12	BZX55C12	BZX85C12	BZX85C12	BZX70C12	BZT03C12	1N5349B	BZW03-C12	BZY93C12#	BZY91C12
13V	BZX79C13	BZX55C13	BZX85C13	BZX85C13	BZX70C13	BZT03C13	1N5350B		BZY93C13#	
15V	BZX79C15	BZX55C15	BZX85C15	BZX85C15	BZX70C15	BZT03C15	1N5352B		BZY93C15#	BZY91C15#

16V	BZX79C16	BZX55C16	BZX85C16	BZV85C16	BZX70C16	BZT03C16	1N5353B		BZY93C16#	
18V	BZX79C18	BZX55C18	BZX85C18	BZV85C18	BZX70C18	BZT03C18	1N5355B		BZY93C18#	BZY91C18#
20V	BZX79C20	BZX55C20	BZX85C20	BZV85C20	BZX70C20	BZT03C20	1N5357B		BZY93C20#	BZY91C20
22V	BZX79C22	BZX55C22	BZX85C22	BZV85C22	BZX70C22	BZT03C22	1N5358B		BZY93C22	
24V	BZX79C24	BZX55C24	BZX85C24	BZV85C24	BZX70C24	BZT03C24	1N5359B	BZW03-C24 NEW	BZY93C24#	BZY91C24
27V	BZX79C27	BZX55C27	BZX85C27	BZV85C27	BZX70C27	BZT03C27	1N5361B	BZW03-C27 NEW	BZY93C27#	BZY91C27
30V	BZX79C30	BZX55C30	BZX85C30	BZV85C30	BZX70C30	BZT03C30	1N5363B		BZY93C30#	BZY91C30
33V	BZX79C33	BZX55C33	BZX85C33	BZV85C33	BZX70C33	BZT03C33	1N5364B		BZY93C33#	BZY91C33
36V	BZX79C36	BZX55C36	BZX85C36	BZV85C36	BZX70C36	BZT03C36	1N5365B	BZW03-C36 NEW	BZY93C36	BZY91C36
39V	BZX79C39	BZX55C39	BZX85C39	BZV85C39	BZX70C39	BZT03C39	1N5366B		BZY93C39#	
43V	BZX79C43	BZX55C43	BZX85C43	BZV85C43	BZX70C43	BZT03C43	1N5367B		BZY93C43	BZY91C43
47V	BZX79C47	BZX55C47	BZX85C47	BZV85C47	BZX70C47	BZT03C47	1N5368B	BZW03-C47 NEW	BZY93C47	BZY91C47
51V	BZX79C51	BZX55C51	BZX85C51	BZV85C51	BZX70C51	BZT03C51	1N5369B	BZW03-C51 NEW	BZY93C51	BZY91C51
56V	BZX79C56	BZX55C56	BZX85C56	BZV85C56	BZX70C56	BZT03C56	1N5370B		BZY93C56	
62V	BZX79C62	BZX55C62	BZX85C62	BZV85C62	BZX70C62	BZT03C62	1N5372B		BZY93C62	BZY91C62
68V	BZX79C68	BZX55C68	BZX85C68	BZV85C68	BZX70C68	BZT03C68	1N5373B		BZY93C68	BZY91C68
75V	BZX79C75	BZX55C75	BZX85C75	BZV85C75	BZX70C75	BZT03C75	1N5374B	BZW03-C75 NEW	BZY93C75#	BZY91C75
82V		BZX55C82	BZX85C82			BZT03C82	1N5375B	BZW03-C82 NEW		
91V		BZX55C91	BZX85C91			BZT03C91	1N5377B			
100V		BZX55C100	BZX85C100			BZT03C100	1N5378B			
110V		BZX55C110	BZX85C110			BZT03C110	1N5379B			
120V		BZX55C120	BZX85C120			BZT03C120	1N5380B			
130V		BZX55C130	BZX85C130			BZT03C130	1N5381B			
150V		BZX55C150	BZX85C150			BZT03C150	1N5383B			
160V		BZX55C160	BZX85C160			BZT03C160	1N5384B			
180V		BZX55C180	BZX85C180			BZT03C180	1N5386B			
200V		BZX55C200	BZX85C200			BZT03C200	1N5388B			
220V						BZT03C220				
240V						BZT03C240				
270V						BZT03C270				

Transistors (1)

type	material	case	application	P_T	I_C	V_{CEO}	V_{CBO}	h_{FE}	f_T (typ)
AC127	NPN Ge	TO1	Audio output	340 mW	500 mA	12 V	32 V	50	2·5 MHz
AC128	PNP Ge	TO1	Audio output	700 mW	−1 A	−16 V	−32 V	60–175	1·5 MHz
AD149	PNP Ge	TO3(A)	Audio output	*22·5 W at 50 °C	−3·5 A	−50 V	−50 V	30–100	0·5 MHz
AD161 } Pair	NPN Ge	SO55(A)	Audio matched pair	*4 W at 72 °C	3 A	20 V	32 V	50–300	3 MHz
AD162 }	PNP Ge	SO55(A)		*6 W at 63 °C	−3 A	−20 V	−32 V	50–300	1·5 MHz
AF127	PNP Ge	TO72(A)	I.F. Applications	60 mW	−10 mA	−20 V	−20 V	150 ◆	75 MHz
BC107	NPN Si	TO18	Audio driver stages (complement BC177)	360 mW	100 mA	45 V	50 V	110–450	250 MHz
BC108	NPN Si	TO18	General purpose (complement BC178)	360 mW	100 mA	20 V	30 V	110–800	250 MHz
BC109	NPN Si	TO18	Low noise audio (complement BC179)	360 mW	100 mA	20 V	30 V	200–800	250 MHz
BC142	NPN Si	TO39	Audio driver	800 mW	800 mA	60 V	80 V	20 (min.)	80 MHz
BC143	PNP Si	TO39	Audio driver	800 mW	−800 mA	−60 V	−60 V	25 (min.)	160 MHz
BC177	PNP Si	TO18	Audio driver stages (complement BC107)	300 mW	−100 mA	−45 V	−50 V	125–500	200 MHz
BC178	PNP Si	TO18	General purpose (complement BC108)	300 mW	−100 mA	−25 V	−30 V	125–500	200 MHz
BC 179	PNP Si	TO18	Low Noise Audio (complement BC109)	300 mW	−100 mA	−20 V	−25 V	240–500	200 MHz
BC182L	NPN Si	TO92(A)	General purpose	300 mW	200 mA	50 V	60 V	100–480	150 MHz
BC183L	NPN Si	TO92(A)	General purpose (complement BC213L)	300 mW	200 mA	30 V	45 V	100–850	280 MHz
BC184L	NPN Si	TO92(A)	General purpose	300 mW	200 mA	30 V	45 V	250 (min.)	150 MHz
BC212L	PNP Si	TO92(A)	General purpose	300 mW	−200 mA	−50 V	−60 V	60–300	200 MHz
BC213L	PNP Si	TO92(A)	General purpose (complement BC183L)	300 mW	−200 mA	−30 V	−45 V	80–400	350 MHz

Type	Material	Case	Application	Power	I_C	V	V	h_{FE}	f_T
BC214L	PNP Si	TO92(A)	General purpose	300 mW	−200 mA	−30 V	−45 V	140–600	200 MHz
BC237B	NPN Si	TO92(B)	Amplifier	350 mW	100 mA	45 V	—	120–800	100 MHz
BC307B	PNP Si	TO92(B)	Amplifier	350 mW	100 mA	45 V	50 V	120–800	280 MHz
BC327	PNP Si	TO92(B)	Driver	625 mW	−500 mA	−45 V	−50 V	100–600	260 MHz
BC337	NPN Si	TO92(B)	Audio driver	625 mW at 45 °C	500 mA	45 V	50 V	100–600	200 MHz
BC441	NPN Si	TO39	General purpose (complement BC461)	1 W	2 A peak	60 V	75 V	40–250	50 MHz (min.)
BC461	PNP Si	TO39	General purpose (complement BC441)	1 W	−2 A peak	−60 V	−75 V	40–250	50 MHz
BC477	PNP Si	TO18	Audio driver stages	360 mW	−150 mA	−80 V	−80 V	110–950	150 MHz
BC478	PNP Si	TO18	General purpose	360 mW	−150 mA	−40 V	−40 V	110–800	150 MHz
BC479	PNP Si	TO18	Low noise audio amp.	360 mW	−150 mA	−40 V	−40 V	110–800	150 MHz
BCY70	PNP Si	TO18	General purpose	360 mW	−200 mA	−40 V	−50 V	150	200 MHz
BCY71	PNP Si	TO18	Low noise general purpose	360 mW	−200 mA	−45 V	−45 V	100–400	200 MHz
BD131	NPN Si	TO126[m]	General purpose – medium power	*15 W at 60 °C	3 A	45 V	70 V	20 (min.)	60 MHz
BD132	PNP Si	TO126[m]	General purpose – medium power	*15 W at 60 °C	−3 A	−45 V	−45 V	20 (min.)	60 MHz
BD131 ⎫ / BD132 ⎭ Pair	NPN Si / PNP Si	TO126[m]	Audio matched pair	*15 W at 60 °C	3 A / −3 A	45 V / −45 V	70 V / −45 V	20 (min.)	60 MHz
BD135	NPN Si	SOT32[m]	Audio driver	*12·5 W at 25 °C	1·5 A	45 V	45 V	40–250	50 MHz
BD136	PNP Si	SOT32[m]	Audio driver	*12·5 W at 25 °C	−1·5 A	−45 V	−45 V	40–250	75 MHz
BD437 ⎫ / BD438 ⎭	NPN Si / PNP Si	TO126[m]	Power switching complementary	*36 W at 25 °C	4 A	45 V	45 V	40	3 MHz

Transistors (2)

BCY70	PNP Si	TO18	General purpose	360 mW	−200 mA	−40 V	−50 V	150	200 MHz
BCY71	PNP Si	TO18	Low noise general purpose	360 mW	−200 mA	−45 V	−45 V	100–400	200 MHz
BD131	NPN Si	TO126 [m]	General purpose – medium power	*15 W at 60 °C	3 A	45 V	70 V	20 (min.)	60 MHz
BD132	PNP Si	TO126 [m]	General purpose – medium power	*15 W at 60 °C	−3 A	−45 V	−45 V	20 (min.)	60 MHz
BD131 } Pair	NPN Si	} TO126 [m]	Audio matched pair	*15 W at 60 °C	3 A	45 V	70 V	20 (min.)	60 MHz
BD132	PNP Si			*15 W at 60 °C	−3 A	−45 V	−45 V	20 (min.)	60 MHz
BD135	NPN Si	SOT32 [m]	Audio driver	*12·5 W at 25 °C	1·5 A	45 V	45 V	40–250	50 MHz
BD136	PNP Si	SOT32 [m]	Audio driver	*12·5 W at 25 °C	−1·5 A	−45 V	−45 V	40–250	75 MHz
BD437	NPN Si	} TO126 [m]	Power switching complementary	*36 W at 25 °C	4 A	45 V	45 V	40	3 MHz
BD438	PNP Si								
BD679	NPN Si	} TO126 [m]	Audio complementary Darlington	*40 W at 25 °C	6 A	80 V	100 V	2200	60 kHz
BD680	PNP Si			*40 W at 25 °C	6 A	80 V	80 V	2200	60 kHz
BDX33C	NPN Si	} TO220(A) [nn] (A)	Power switching Darlington	*70 W at 25 °C	10 A	100 V	100 V	750 (min.)	—
BDX34C	PNP Si								
BF259	NPN Si	TO39	High-voltage video amplifier	800 mW	100 mA	300 V	300 V	25	90 MHz
BF337	NPN Si	TO39	Video amplifier	*3 W at 125 °C	100 mA	200 V	250 V	20 (min.)	80 MHz

Device	Type	Case	Application	Power	I_C	V	V	h_{FE}	f_T
BFY50	NPN Si	TO39	High voltage general purpose	800 mW	1 A	35 V	80 V	30	60 MHz
BFY51	NPN Si	TO39	General purpose	800 mW	1 A	30 V	60 V	40	50 MHz
BFY52	NPN Si	TO39	General purpose	800 mW	1 A	20 V	40 V	60	50 MHz
TIP31A	NPN Si	TO220^m(A)	Plastic medium power complementary	*40 W at 25 °C	3 A	60 V	60 V	10–60	8 MHz
TIP32A	PNP Si							10–40	
TIP31C	NPN Si	TO220(A)	Plastic medium power complementary	*40 W at 25 °C	3 A	100 V	100 V	10–50	8 MHz
TIP32C	PNP Si								
TIP33A	NPN Si	TAB^m(A)	Audio output complementary	*80 W at 25 °C	10 A	60 V	60 V	20–100	3 MHz
TIP34A	PNP Si								
2N2905	PNP Si	TO5(A)	Switching	600 mW	–600 mA	–40 V	–60 V	100–300	200 MHz (min.)
2N2905A	PNP Si	TO5(A)	Switching	600 mW	–600 mA	–60 V	–60 V	100–300	200 MHz (min.)
2N2907A	PNP Si	TO18(A)	Switching	400 mW	–600 mA	–60 V	–60 V	100–300	200 MHz (min.)
2N3019	NPN Si	TO39(A)	General purpose	500 mW	1 A	80 V	140 V	90 min.	100 MHz
2N3053	NPN Si	TO39(A)	General purpose	800 mW	1 A	40 V	60 V	50–250	100 MHz
2N3055E	NPN Si	TO3(A)	High power epitaxial (complement MJ2955)	*115 W at 25 °C	15 A	60 V	100 V	20–70	2·5 MHz
2N3055H	NPN Si	TO3(A)	High power homotaxial (complement PNP3055)	*115 W at 25 °C	15 A	60 V	100 V	20–70	1 MHz
PNP3055	PNP Si	TO3(A)	High power (complement 2N3055H)	*115 W at 25 °C	–15 A	–60 V	–100 V	20–70	0·8 MHz
2N3440	NPN Si	TO39(A)	General purpose	1 W	1 A	250 V	300 V	40–160	15 MHz
2N3702	PNP Si	TO92(A)	General purpose	360 mW	–200 mA	–25 V	–40 V	60–300	100 MHz

Transistor pin connections

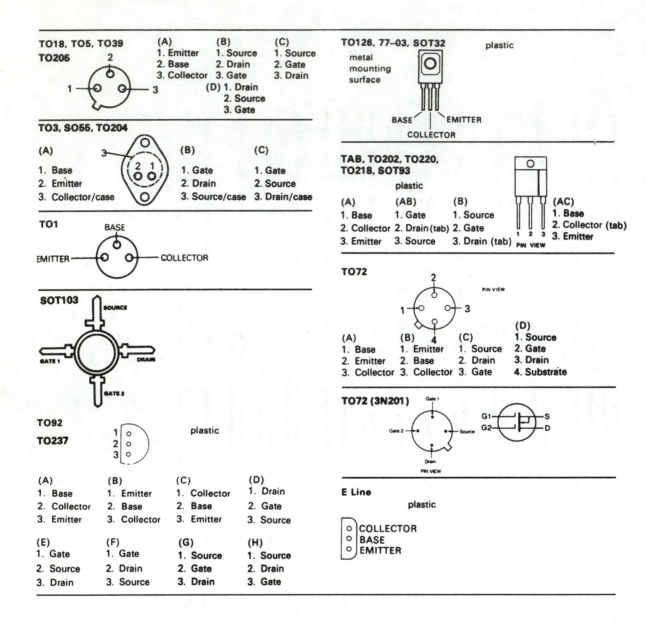

TO18, TO5, TO39
TO205

(A)
1. Emitter
2. Base
3. Collector

(B)
1. Source
2. Drain
3. Gate

(C)
1. Source
2. Gate
3. Drain

(D) 1. Drain
2. Source
3. Gate

TO3, SO55, TO204

(A)
1. Base
2. Emitter
3. Collector/case

(B)
1. Gate
2. Drain
3. Source/case

(C)
1. Gate
2. Source
3. Drain/case

TO1

BASE
EMITTER
COLLECTOR

SOT103

SOURCE
GATE 1
DRAIN
GATE 2

TO92
TO237

plastic

1
2
3

(A)
1. Base
2. Collector
3. Emitter

(B)
1. Emitter
2. Base
3. Collector

(C)
1. Collector
2. Base
3. Emitter

(D)
1. Drain
2. Gate
3. Source

(E)
1. Gate
2. Source
3. Drain

(F)
1. Gate
2. Drain
3. Source

(G)
1. Source
2. Gate
3. Drain

(H)
1. Source
2. Drain
3. Gate

TO126, 77–03, SOT32   plastic

metal
mounting
surface

BASE   EMITTER
COLLECTOR

TAB, TO202, TO220,
TO218, SOT93

plastic

(A)
1. Base
2. Collector
3. Emitter

(AB)
1. Gate
2. Drain (tab)
3. Source

(B)
1. Source
2. Gate
3. Drain (tab)

(AC)
1. Base
2. Collector (tab)
3. Emitter

1   2   3
PIN VIEW

TO72

2
1   3
4
PIN VIEW

(A)
1. Base
2. Emitter
3. Collector

(B)
1. Emitter
2. Base
3. Collector

(C)
1. Source
2. Drain
3. Gate

(D)
1. Source
2. Gate
3. Drain
4. Substrate

TO72 (3N201)

Gate 1
Gate 2   Source
Drain
PIN VIEW

G1   S
G2   D

E Line

plastic

COLLECTOR
BASE
EMITTER

Field effect transistors (1)

JUNCTION FETs

*Package and Pin Connection	Channel	V_{DS} max V	V_{GS} (Off) max V	I_G (I_D*) max mA	P_{tot} @25°C mW	I_{DSS} mA	GFS mA/V	App'n Code	Order Code
TO-92 (d)	N	30	8	10	200	2—6.5	3—6.5	●	★ BF244A
TO-92 (d)	N	30	8	10	200	6—15	3—6.5	●	BF244B
TO-92 (e)	N	30	8	10	200	2—6.5	3—6.5	●	★ BF245A
TO-92 (e)	N	30	8	10	200	6—15	3—6.5	●	BF245B
TO-92 (e)	N	30	8	10	200	12—25	3—6.5	●	BF245C
TO-92 (f)	N	30	7.5	10	300	3—7	4.5 min	●	BF256A
TO-92 (f)	N	30	7.5	10	300	11—18	4.5 min	●	BF256C
TO-72 (c)	N	30	8	10	300	8—20	3.5 min	●	BFW10
TO-72 (c)	N	30	6	10	300	4—10	3 min	●	BFW11
TO-18 (b)	N	40	5	50	350	10 min	—	◆	BSV80
TO-92 (b)	N	25	4.5	50	360	100 min	—	◇	J107
TO-92 (b)	N	25	10	50	360	80 min	—	◇	J108
TO-92 (b)	N	25	6	50	360	40 min	—	◇	J109
TO-92 (b)	N	25	4	50	360	10 min	—	◇	J110
TO-92 (b)	N	35	10	50	360	20 min	—	◇	J111
TO-92 (b)	N	35	5	50	360	5 min	—	◇	★ J112
TO-92 (b)	N	35	3	50	360	2 min	—	◇	J113
TO-92 (a)	P	30	10	50	360	20—125	—	◇	J174
TO-92 (a)	P	30	6	50	360	7—70	—	◇	★ J175
TO-92 (a)	P	30	4	50	360	2—35	—	◇	J176
TO-92 (a)	P	30	2.25	50	360	1.5—20	—	◇	★ J177
TO-92 (b)	N	40	1.5	50	360	0.2—1	0.5 min	○	J201
TO-92 (b)	N	40	4	50	360	0.9—4.5	1 min	○	J202
TO-92 (b)	N	25	3	10	360	2—15	4—12	●	J210
TO-92 (b)	N	25	4.5	10	360	7—20	6—12	●	J211
TO-92 (b)	N	40	5	50	360	2—6	1.5—4	△	J231
TO-92 (a)	P	30	2	50	360	2—15	6—15	●	J270
TO-92 (b)	N	25	7	10	360	4—45	4.5—9	●	J300
TO-92 (b)	N	30	6	10	360	5—15	4.5—7.5	●	J304
TO-92 (b)	N	30	3	10	360	1—8	3 min	●	J305
TO-92 (b)	N	25	4	10	360	12—30	10 min	●	★ J309
TO-92 (b)	N	25	6.5	10	360	24—60	8 min	●	★ J310
TO-92 (b)	N	25	8	10	360	2—20	2—7.5	●	MPF102
TO-92 (c)	N	25	6	20	500	24—60	10—18	●	U310
TO-18 (b)	N	40	10	10	360	30 min	—	◇	U1897

Field effect transistors (*contd*)

JUNCTION FETs

Package	Channel								Type
TO-92 (b)	N	40	7	10	360	15 min	—	◊	U###
TO-18 (b)	N	15	7	10	300	—	—	V	VCR2N
TO-18 (a)	P	15	7	10	300	R_{ds} (on) = 200Ω max		V	VCR3P
TO-18 (b)	N	15	7	10	300	R_{ds} (on) = 600Ω max		V	VCR4N
TO-72 (d)	N	15	5	10	300	R_{ds} (on) = 8KΩ max		V	VCR7N
TO-92 (d)	N	25	8	10	200	2—20	2—6.5	○	★ 2N3819
TO-92 (a)	P	20	8	15*	360	0.3—15	0.8—5	○	★ 2N3820
TO-72 (c)	N	50	4	2.5*	300	0.5—2.5	1.5—4.5	○	2N3821
TO-72 (c)	N	50	6	10	300	2—10	3—6.5	○	2N3822
TO-72 (c)	N	50	8	10	300	4—20	3.5—6.5	●	2N3823
TO-72 (c)	N	50	—	10	300	—	—	♦	2N3824
TO-18 (b)	N	40	10	150*	360	30 min	—	♦	2N4091
TO-18 (b)	N	40	7	150*	360	15 min	—	♦	2N4092
TO-18 (b)	N	40	5	150*	360	8 min	—	♦	2N4093
TO-72 (c)	N	40	1.8	50	300	0.03—0.09	0.07—0.21	○	2N4117A
TO-72 (c)	N	40	3	50	300	0.08—0.24	0.08—0.25	○	★ 2N4118
TO-72 (d)	N	30	4	10	300	0.5—3	1—4	●	★ 2N4220
TO-18 (b)	N	50	1	50	300	0.2—0.6	0.6—1.8	●	2N4338
TO-18 (b)	N	40	10	—	300	50—150	—	♦	2N4391
TO-18 (b)	N	40	5	—	300	25—75	—	♦	2N4392
TO-18 (b)	N	40	3	—	300	5—30	—	♦	2N4393
TO-72 (c)	N	30	6	10	300	5—15	4.5—7.5	●	★ 2N4416
TO-72 (c)	N	35	6	10	300	5—15	4.5—7.5	●	2N4416A
TO-18 (b)	N	40	10	50	1800	50 min	—	♦	2N4856
TO-18 (b)	N	40	6	50	1800	20—100	—	♦	2N4857
TO-18 (b)	N	40	4	50	1800	8—80	—	♦	★ 2N4858
TO-18 (b)	N	30	10	50	1800	50 min	—	♦	2N4859A
TO-18 (b)	N	30	6	50	1800	20—100	—	♦	2N4860
TO-18 (a)	P	30	4	50	500	5—25	—	♦	2N5116
TO-18 (b)	N	25	4	400*	300	30 min	—	♦	2N5434
TO-92 (b)	N	25	0.5	16*	310	1—5	2—5	○	2N5457
TO-92 (b)	N	25	1	16*	310	2—9	1.5—5.5	○	2N5458
TO-92 (b)	N	25	2	16*	310	4—16	2—6	○	2N5459
TO-92 (c)	P	40	6	16*	310	1—5	1—4	△	2N5460
TO-92 (c)	P	40	7.5	16*	310	2—9	1.5—5	△	2N5461
TO-92 (c)	P	40	9	16*	310	4—16	2—6	△	2N5462
TO-92 (b)	N	25	3	10	360	1—5	3—6	●	2N5484
TO-92 (b)	N	25	4	10	360	4—10	3.5—7	●	2N5485
TO-92 (b)	N	25	6	10	360	8—20	4—8	●	★ 2N5486
TO-92 (b)	N	30	8	250*	310	25 min	—	♦	2N5639

Field effect transistors (2)

JUNCTIONS FETS

type	channel mtl.	case	application	P_T	V_P	V_{DG}	V_{DS}	I_{DSS} (min.)	I_{GSS}	Y_{FS} (min.)	C_{iss}
BF244A	n	TO92(D)	Wide band amplifier	300 mW	2·2 V	30 V	30 V	2 mA	5 nA	3000 μS*	6 pF
BF245A	n	TO92(E)	V/UHF amplifier	360 mW	—	30 V	30 V	2 mA	5 nA	3 to 6·5 mS	3 pF
BF981	n	SOT 103	VHF tuner, etc	225 mW	2·5 V	—	20 V	4 mA	50 nA	14 mS	2·1 pF
BS107	n	TO92(G)	Switching	600 mW	—	—	200 V	30 nA	10 pA	—	72 pF
BS170	n	TO92(G)	Switching	830 mW	—	—	60 V	—	10 pA	—	60 pF
2N3819	n	TO92(D)	General purpose amplifier	200 mW	8 V	25 V	25 V	2 mA	2 nA	2000 μS*	8 pF
2N3820	p	TO92(D)	General purpose amplifier	200 mW	8 V	−20 V	−20 V	0·3 mA	20 nA	800 μS*	32 pF
2N4092	n	TO18(B)	Switching	1·8 W	—	40 V	40 V	15 mA	—	—	—
2N4118	n	TO72(D)	General purpose amplifier	300 mW	−3 V	40 V	—	0·08 mA	10 pA	—	3 pF
2N4220	n	TO72(C)	Low noise	300 mW	—	30 V	30 V	3 mA	100 pA	1000 to 4000 μS*	4·5 pF
2N4351	n	TO72(D)	Switching	300 mW	—	35 V	25 V	30 mA	10 pA	1000 μS	5·5 pF
2N4391	n	TO18(B)	J-F.E.T. switch	1·8 W	−10 V	40 V	40 V	50 mA	100 pA	—	14 pF
2N4392	n	TO18(B)	Switching	1·8 W	—	40 V	40 V	25 mA	100 pA	—	14 pF
2N4393	n	TO18(B)	Switching	1·8 W	—	40 V	40 V	5 mA	100 pA	—	14 pF
2N4416	n	TO72(C)	V/UHF amplifier	300 mW	—	35 V	30 V	5 mA	100 pA	4000 μS	4·0 pF
2N4858	n	TO18(D)	J-F.E.T. switch	1·8 W	−4 V	40 V	40 V	8 mA	0·5 μA	—	18 pF
2N4858A	n	TO18(D)	Lower noise	1·8 W	−4 V	40 V	40 V	8 mA	0·5 μA	—	18 pF
2N4859A	n	TO18(D)	J-F.E.T. switch	1·8 W	−10 V	30 V	30 V	50 mA	0·5 μA	—	18 pF
2N4860A	n	TO18(D)	J-F.E.T. switch	1·8 W	−6 V	30 V	30 V	20 mA	0·5 μA	—	18 pF
2N4861	n	TO18(C)	Switching	360 mW	—	30 V	30 V	8 mA	250 pA	—	10 pF
2N5457	n	TO92(E)	General purpose amplifier	310 mW	6 V	25 V	25 V	1 mA	1 nA	1000 μS*	4·5 pF (typ.)
2N5460	p	TO92(F)	General purpose amplifier	310 mW	6 V	−40 V	−40 V	1 mA	5 nA	1000 μS*	5 pF (typ.)
2N5461	p	TO92(E)	Amplifier	310 mW	—	40 V	—	2 mA	500 pA	1500 to 5000 μS*	5 pF
2N5486	n	TO92(E)	V/UHF	310 mW	—	25 V	—	8 mA	1 nA	—	5 pF

Field effect transistors (2) (*contd*)

POWER MOSFETS

type	channel mtl.	case	P_T	R_{DS} (max.)	I_D (max.)	V_{DG}	V_{DS}	(TH) (max.)	I_{DSS} (max.)	I_{GSS} (max.)	tr, tf (max.)	gfs* (min.)
BUZ11	n	TO220(AB)	75 W	0·04 Ω	30 A	50 V	50 V	4 V	250 μA	100 nA	110, 170 ns	4 S
IRF120	n	TO204(C)	40 W	0·3 Ω	6 A	100 V	100 V	4 V	1 mA	100 nA	70 ns	1·5 S
IRF130	n	TO204(C)	75 W	0·18 Ω	12 A	100 V	100 V	4 V	1 mA	100 nA	150 ns	3 S
IRF330	n	TO204(C)	75 W	1·0 Ω	4 A	400 V	400 V	4 V	1 mA	100 nA	100 ns	2 S
IRF510	n	TO220(AB)	20 W	0·6 Ω	3 A	100 V	100 V	4 V	0·5 mA	500 nA	25 ns	1 S
IRF511	n	TO220(AB)	20 W	0·6 Ω	4 A	60 V	60 V	4 V	500 nA	500 nA	25/20 ns	1·0 S
IRF520	n	TO220(AB)	40 W	0·3 Ω	5 A	100 V	100 V	4 V	1 mA	500 nA	70 ns	1·5 S
IRF530	n	TO220(AB)	75 W	0·18 Ω	10 A	100 V	100 V	4 V	1 mA	500 nA	150 ns	3 S
IRF531	n	TO220(AB)	75 W	0·18 Ω	14 A	60 V	60 V	4 V	1000 μA	500 nA	75/45 ns	4·0 S
IRF610	n	TO220(AB)	20 W	1·5 Ω	2 A	200 V	200 V	4 V	0·5 mA	500 nA	25 ns	0·8 S
IRF620	n	TO220(AB)	40 W	0·8 Ω	5 A	200 V	200 V	4 V	1000 μA	500 nA	60/60 ns	1·3 S
IRF621	n	TO220(AB)	40 W	0·8 Ω	5 A	150 V	150 V	4 V	1 mA	500 nA	60 ns	1·3 S
IRF630	n	TO220(AB)	75 W	0·4 Ω	6 A	200 V	200 V	4 V	1 mA	500 nA	140 ns	2·5 S
IRF640	n	TO220(AB)	125 W	0·18 Ω	11 A	200 V	200 V	4 V	1 mA	500 nA	60 ns	6 S
IRF643	n	TO220(AB)	125 W	0·22 Ω	16 A	150 V	150 V	4 V	1 mA	500 nA	60 ns	6 S
IRF720	n	TO220(AB)	40 W	1·8 Ω	2·5 A	400 V	400 V	4 V	1 mA	500 nA	50 ns	1 S
IRF730	n	TO220(AB)	75 W	1·0 Ω	3·5 A	400 V	400 V	4 V	1 mA	500 nA	100 ns	2 S
IRF822	n	TO220(AB)	40 W	4 Ω	2 A	500 V	500 V	4 V	1 mA	500 nA	50 ns	1 S
IRF830	n	TO220(AB)	75 W	1·3 Ω	3 A	500 V	500 V	4 V	1 mA	500 nA	80 ns	1·5 S
IRF831	n	TO220(AB)	75 W	1·5 Ω	4·5 A	450 V	450 V	4 V	1000 μA	500 nA	30/30 ns	2·5 S
IRF840	n	TO220(AB)	125 W	0·85 Ω	5 A	500 V	500 V	4 V	1 mA	500 nA	30 ns	4 S
IRFD110	n	4 pin d.i.l.	1 W	0·6 Ω	1 A	100 V	100 V	—	—	—	—	—
IRFD111	n	4 pin d.i.l.	1 W	0·6 Ω	1 A	60 V	60 V	—	—	—	—	—
IRFD113	n	4 pin d.i.l.	1 W	0·8 Ω	0·8 A	60 V	60 V	—	—	—	—	—
IRFD123	n	4 pin d.i.l.	1 W	0·4 Ω	1·1 A	60 V	60 V	—	—	—	—	—
IRFF110	n	TO205(C)	15 W	0·6 Ω	3·5 A	100 V	100 V	4 V	1 mA	100 nA	25 ns	1 S
MTP3055A	n	TO220ᵐ(AB)	40 W	0·15 Ω	12 A	60 V	60 V	4·5 V	50 μA	100 nA	65 ns	4·5 S
MTH7N50	n	TO218(AB)	—	—	7 A	500 V	500 V	—	—	—	—	—
MTH8P20	p	TO218(AB)	125 W	0·7 Ω	8 A	200 V	200 V	4·5 V	0·2 mA	100 nA	120, 80 ns	2 S
MTH13N50	n	TO218(AB)	150 W	0·4 Ω	13 A	500 V	500 V	4·5 V	1 mA	500 nA	180/180 ns	5 mS
MTH15N20	n	TO218(AB)	150 W	0·16 Ω	15 A	200 V	200 V	4·5 V	0·25 mA	500 nA	300, 250 ns	4 S
MTM15N50	n	TO204(C)	250 W	0·4 Ω	15 A	500 V	500 V	4·5 V	0·25 mA	500 nA	300, 240 ns	4 S
MTP2N50	n	TO220(AB)	75 W	6 Ω	2 A	500 V	500 V	4·5 V	0·25 mA	500 nA	100, 50 ns	0·5 S
MTP2P50	p	TO220(AB)	75 W	6 Ω	2 A	500 V	500 V	4·5 V	0·25 mA	500 nA	100, 50 ns	0·5 S
MTP10N10	n	TO220(AB)	75 W	0·33 Ω	10 A	100 V	100 V	4·5 V	0·25 mA	500 nA	150, 50 ns	2·5 S
MTP10N10M*	n	TO220(5 pin)	75 W	0·25 Ω	10 A	100 V	100 V	4·5 V	0·2 mA	100 nA	150, 50 ns	2·5 S
MTP25N06	n	TO220(AB)	100 W	0·08 Ω	25 A	60 V	60 V	4·5 V	0·25 mA	500 nA	450, 200 ns	6 S

VMOS FETs

type	channel mtl.	case	P_T	R_{DS} (max.)	I_D (max.)	V_{DG}	V_{DS}	V_{GS} (th)	I_{DSS}	I_{GSS}	tr, tf (max.)	gfs (min.)*
VN10LM	n	TO237†	1 W	5 Ω	0·5 A	60 V	60 V	2·5 V max.	10 μA	10 μA	5 ns	100 mS
VN46AF	n	TO202(B)ᵐ	15 W	3 Ω	2 A	40 V	40 V	1·7 V typ.	10 μA	10 nA	5 ns	150 mS
VN66AF	n	TO202(B)ᵐ	15 W	3 Ω	2 A	60 V	60 V	1·7 V typ.	10 μA	10 nA	5 ns	150 mS
VN88AF	n	TO202(B)ᵐ	15 W	4 Ω	2 A	80 V	80 V	1·7 V typ.	10 μA	10 nA	5 ns	150 mS

Voltage regulators (1)

FIXED VOLTAGE REGULATORS

Standard

100mA

3 Lead TO-92

3 Lead TO-5

Voltage	Mftr.	Pins/Package	Special Features	Order Code
+ 5V	NSC	3/TO-99	....	LM78L05ACH
+ 5V	NSC	■8	....	LM78L05ACM
+ 5V	NSC	3/TO-92	....	★ LM78L05ACZ
+ 5V	NSC	3/TO-92	....	LM340LAZ-5
+ 5V	NSC	3/TO-5	....	LM340LAH-5
+ 5V	ST	16/DIL	....	TEA7105DP
+ 6.2V	NSC	3/TO-92	....	UA78L62AWC
+ 8.2V	NSC	3/TO-92	....	UA78L82AWC
+ 9V	NSC	3/TO-92	....	UA78L09AWC
+ 12V	NSC	■8	....	LM78L12ACM
+ 12V	NSC	3/TO-92	....	★ LM78L12ACZ
+ 12V	NSC	3/TO-92	....	LM340LAZ-12
+ 12V	NSC	3/TO-5	....	LM340LAH-12
+ 15V	NSC	■8	....	LM78L15ACM
+ 15V	NSC	3/TO-92	....	★ LM78L15ACZ
+ 15V	NSC	3/TO-92	....	LM340LAZ-15
− 5V	NSC	■8	....	LM79L05ACM
− 5V	NSC	3/TO-92	....	★ LM79L05ACZ
− 12V	NSC	■8	....	LM79L12ACM
− 12V	NSC	3/TO-92	....	★ LM79L12ACZ
− 15V	NSC	■8	....	LM79L15ACM
− 15V	NSC	3/TO-92	....	★ LM79L15ACZ

200mA

3 Lead TO-99

Voltage	Mftr.	Pins/Package	Special Features	Order Code
+ 5V	NSC	3/TO-202	....	LM342P-5
+ 5V	NSC	3/TO-5	....	LM140LAH-5
+ 12V	NSC	3/TO-202	....	LM342P-12
+ 12V	NSC	3/TO-99	....	LM78L12ACH
+ 12V	NSC	3/TO-5	....	LM140LAH-12
+ 15V	NSC	3/TO-202	....	LM342P-15
+ 15V	NSC	3/TO-99	....	LM78L15ACH

■ Including Surface Mount Devices

Voltage regulators (1) (*contd*)

3 pin TO-202	500mA	+5V	NSC	3/TO-202	....	LM341P-5
		+5V	ST	3/TO-220	....	L78M05CV
		+5V	NSC	3/TO-220	....	LM78M05CT
		+5V	TI	3/TO-220	....	UA78M05CKC
		+5V	NSC	3/TO-5	....	LM309H
		+12V	NSC	3/TO-202	....	LM341P-12
		+12V	ST	3/TO-220	....	L78M12CV
		+12V	NSC	3/TO-220	....	LM78M12CT
		+12V	TI	3/TO-220	....	UA78M12CKC
		+15V	NSC	3/TO-202	....	LM341P-15
		+15V	ST	3/TO-220	....	L78M15CV
		+15V	NSC	3/TO-220	....	LM78M15CT
		+15V	TI	3/TO-220	....	UA78M15CKC
3 pin TO-220		+24V	ST	3/TO-220	....	L78M24CV
		−5V	NSC	3/TO-202	....	LM79M05CP
		−5V	TI	3/TO-220	....	UA79M05CKC
		−12V	NSC	3/TO-202	....	LM79M12CP
		−12V	NSC	3/TO-202	....	LM320MP-12
		−12V	TI	3/TO-220	....	UA79M12CKC
		−15V	NSC	3/TO-202	....	LM79M15CP
		−15V	NSC	3/TO-202	....	LM320MP-15
		−15V	TI	3/TO-220	....	UA79M15CKC
2 pin TO-3	1.5A	+5V	ST	3/TO-220	....	L7805CV
		+5V	NSC	3/TO-220	....	★ LM7805CT
		+5V	NSC	3/TO-220	100% burn in....................	LM340T-5
		+5V	NSC	3/TO-220	± 2%	LM340AT-5
		+5V	TI	3/TO-220	1%	TL780-05CKC
		+5V	TI	3/TO-220	....	UA7805CKC
		+5V	NSC	2/TO-3	Aluminium case	LM309K-ALU
		+5V	NSC	2/TO-3	Aluminium case	LM340KC-5
		+5V	NSC	2/TO-3	Steel case....................	★ LM309K-STEEL

Voltage regulators (2)

	1.5A	+5V	NSC	2/TO-3	....	**LM340K-5**
		+5V	NSC	2/TO-3	....	**LM109K-STEEL**
		+5V	ST	2/TO-3	....	**UA7805CK**
		+6V	ST	3/TO-220	....	**L7806CV**
		+6V	TI	3/TO-220	....	**UA7806CKC**
		+8V	ST	3/TO-220	....	**L7808CV**
		+8V	ST	2/TO-3	....	**UA7808CK**
3 Lead TO-92		+8V	TI	3/TO-220	....	**UA7808CKC**
		+12V	ST	3/TO-220	....	**L7812CV**
		+12V	NSC	3/TO-220	....	★ **LM7812CT**
		+12V	NSC	3/TO-220	100% Burn in	**LM340T-12**
		+12V	NSC	3/TO-220	±2%	**LM340AT-12**
		+12V	ST	2/TO-3	....	**UA7812CK**
		+12V	TI	3/TO-220	....	**UA7812CKC**
		+12V	NSC	2/TO-3	Aluminium case	**LM340KC-12**
		+12V	TI	TO-220	1%	**TL780-12CKC**
		+12V	NSC	2/TO-3	....	**LM340K-12**
		+15V	ST	3/TO-220	....	**L7815CV**
		+15V	NSC	3/TO-220	....	★ **LM7815CT**
		+15V	NSC	3/TO-220	100% Burn in	**LM340T-15**
		+15V	NSC	3/TO-220	±2%	**LM340AT-15**
		+15V	ST	2/TO-3	....	**UA7815CK**
		+15V	TI	3/TO-220	....	**UA7815CKC**
		+15V	NSC	2/TO-3	Aluminium case	**LM340KC-15**
3 Lead TO-5		+15V	TI	TO-220	1%	**TL780-15CKC**
		+15V	NSC	2/TO-3	....	**LM340K-15**
		+18V	ST	3/TO-220	....	**L7818CV**
		+18V	TI	3/TO-220	....	**UA7818CKC**
		+20V	ST	3/TO-220	....	**L7820CV**
		+24V	ST	3/TO-220	....	**L7824CV**
		+24V	TI	3/TO-220	....	★ **UA7824CKC**
		−5V	ST	3/TO-220	....	**L7905CV**
		−5V	NSC	3/TO-220	....	★ **LM7905CT**
		−5V	NSC	3/TO-220	100% Burn in	**LM320T-5**
		−5V	ST	2/TO-3	....	**UA7905CK**
		−5V	TI	3/TO-220	....	**UA7905CKC**
3 Lead TO-99		−5V	NSC	2/TO-3	....	**LM320K-5**

Voltage regulators (2) (*contd*)

	Current	Voltage	Mfr	Package	Notes	Part
3 pin TO-202	**1.5A** *contd*	−6V	TI	3/TO-220	....	UA7906CKC
		−8V	ST	3/TO-220	....	L7908CV
		−8V	TI	3/TO-220	....	UA7908CKC
		−12V	ST	3/TO-220	....	L7912CV
		−12V	NSC	3/TO-220	....	★ LM7912CT
		−12V	NSC	3/TO-220	100% Burn in	LM320T-12
		−12V	ST	2/TO-3	....	UA7912CK
		−12V	TI	3/TO-220	....	UA7912CKC
		−12V	NSC	2/TO-3	....	LM320K-12
		−15V	ST	3/TO-220	....	L7915CV
		−15V	NSC	3/TO-220	....	★ LM7915CT
		−15V	NSC	3/TO-220	100% Burn in	LM320T-15
		−15V	ST	2/TO-3	....	UA7915CK
		−15V	TI	3/TO-220	....	UA7915CKC
		−15V	NSC	2/TO-3	....	LM320K-15
		−18V	ST	3/TO-220	....	L7918CV
		−18V	TI	3/TO-220	....	UA7918CKC
		−24V	ST	3/TO-220	....	L7924CV
		−24V	TI	3/TO-220	....	★ UA7924CKC
3 pin TO-220	**2.0A**	+5V	ST	3/TO-220	....	★ L78S05CV
		+12V	ST	3/TO-220	....	★ L78S12CV
		+15V	ST	3/TO-220	....	★ L78S15CV
		+18V	ST	3/TO-220	....	L78S18CV
		+24V	ST	3/TO-220	....	★ L78S24CV
	3.0A	+5V	ST	2/TO-3	....	LM323K
		+5V	NSC	2/TO-3	....	LM323K-STEEL
		+5V	MOT	3/TO-220	....	★ MC78T05CT
		+12V	MOT	3/TO-220	....	★ MC78T12CT
		+15V	MOT	3/TO-220	....	★ MC78T15CT
		−5V	NSC	2/TO-3	....	LM345K-5
2 pin TO-3	**5.0A**	+5V	NE	2/TO-204	Thermally protected..............	★ UA78H05ASC
		+12V	NE	2/TO-204	Thermally protected..............	★ UA78H12ASC

Adjustable voltage regulators

	Voltage Range	Mftr.	Pins/Package	Special Features	Order Code
CMOS 4µA Current Drain	+1.6 to +16V	GE-RCA	■8	....	NEW ICL7663SCBA
	+1.6 to +16V	GE-RCA	8/Cerdip	Improved ICL7663BCJA.	ICL7663SCJA
	+1.6 to +16V	GE-RCA	8/DIL	Improved ICL7663BCPA.	★ ICL7663SCPA
8 pin DIL / 6 Lead TO-99					
25mA to 45mA	−.015 to −40V	NSC	10/TO-100	25mA	LM304H
	+4.5V to +40V	NSC	8/TO-99	45mA	LM305H
	+4.5V to +40V	NSC	8/TO-99	45mA	LM305AH
8 pin DIL / 8 Lead TO-100					
100mA to 400mA	+1.2V to +37V	NSC	■8	100mA, ...	LM317LM
	+1.2V to +37V	NSC	3/TO-92	100mA	★ LM317LZ
	−1.2V to −37V	NSC	3/TO-92	100mA	★ LM337LZ
	+1.24V to +29V	NSC	8/DIL	100mA, ±0.5%	LP2951ACN
	+1.24V to +29V	NSC	■8	100mA	LP2951CM
3 Lead TO-92 / 8 Lead TO-100	+1.24V to +29V	NSC	8/DIL	100mA, ±1%	★ LP2951CN
	+3V to +36V	TI	3/TO-92	100mA	★ TL431CLP
	+3V to +36V	MOT	8/DIL	100mA	NEW TL431CP
	+3V to +36V	NSC	3/TO-92	100mA	UA431AWC
	+2V to +37V	NSC	■14	150mA	LM723CM
	+2V to +37V	NSC	14/DIL	150mA	★ LM723CN
	+2V to +37V	ST	14/DIL	150mA	LM723CN-SGS
	+2V to +37V	PC	14/DIL	150mA	UA723CN
	+2V to +37V	NSC	10/TO-100	150mA	★ LM723CH
	+2V to +37V	NSC	14/DIL	150mA	LM723CJ
14 pin DIL / 3 pin TO-220	+2V to +37V	PC	14/DIL	150mA	UA723F
	+2V to +37V	NSC	10/TO-100	150mA −55°C to +125°C	LM723H
	+2V to +37V	NSC	14/DIL	150mA −55°C to +125°C	LM723J
	+3V to +23V	NSC	5/TO-220	150mA	★ LM2931CT
	+1.25V to +20V	ST	5/TO-220	400mA	L4920
500mA to 750mA 3 Lead TO-5	+1.2V to +37V	NSC	3/TO-202	500mA	★ LM317MP
	+1.25V to +125V	TI	3/TO-220	15 to 750mA high voltage	★ TL783CKC
	+1.2V to +37V	NSC	3/TO-5	500mA	LM317H
3 pin TO-202 / 5 pin TO-220					

■ Surface Mount Device

Adjustable voltage regulators (*contd*)

1.0A	+ 1.2V to + 37V	NSC	3/TO-5	500mA	**LM117H**
	− 1.2V to − 37V	NSC	3/TO-202	500mA	**LM337MP**
	− 1.2V to − 37V	NSC	3/TO-5	500mA	**LM337H**
	− 2.2V to − 30V	NSC	4/TO-202	....	**UA79GU1C**
	+ 5V to + 30V	NSC	4/TO-202	....	**UA78GU1C**
1.5A	+ 1.2V to + 37V	NSC	3/TO-220	....	★ **LM317T**
	+ 1.2V to + 37V	ST	3/TO-220	....	**LM317T-SGS**
	− 1.2V to − 37V	ST	3/TO-220	....	**LM337SP**
	− 1.2V to − 37V	NSC	3/TO-220	....	★ **LM337T**
	+ 1.2V to + 37V	NSC	2/TO-3	....	**LM117K-STEEL**
	+ 1.2V to + 37V	ST	2/TO-3	....	**LM317K**
	+ 1.2V to + 37V	NSC	2/TO-3	....	**LM317KC**
	+ 1.2V to + 37V	NSC	2/TO-3	....	★ **LM317K-STEEL**
	− 1.2V to − 37V	ST	2/TO-3	....	**LM337K**
	− 1.2V to − 37V	NSC	2/TO-3	....	**LM337K-STEEL**
	− 1.2V to − 47V	NSC	2/TO-3	High voltage.	**LM337HVK**
	+ 1.2V to + 57V	NSC	2/TO-3	High voltage.	**LM317HVK-STEEL**
	+ 5.0V to + 40V	ST	16/DIL	....	**L4962**
2.0A to 10.0A	+ 2.8V to + 36V	ST	5/TO-220	2.0A	**L200CH**
	+ 2.8V to + 36V	ST	5/TO-220	2.0A	★ **L200CV**
	+ 1.0V to + 40V	ST	7/SIL	2.5A	**L4960**
	+ 1.2V to + 33V	NSC	3/TO-220	3.0A	**LM350T**
	− 1.2V to − 32V	NSC	3/TO-220	3.0A	**LM333T**
	+ 1.2V to + 33V	NSC	2/TO-3	3.0A	**LM350K-STEEL**
	+ 5.1V to + 40V	ST	15/M-watt	4.0A	★ **L296**
	0V to + 30V	NSC	8/TO-3	5.0A	**LH1605CK**
	+ 1.2V to + 32V	ST	2/TO-3	5.0A	**LM338K**
	+ 1.2V to + 32V	NSC	2/TO-3	5.0A	★ **LM338K-STEEL**
	+ 5V to + 24V	NE	4/TO-204	5.0A Thermally protected	★ **UA78HGASC**
	− 2V to − 24V	NE	4/TO-204	5.0A Thermally protected	★ **UA79HGSC**
	+ 10V to + 60V	NSC	8/TO-3	7.0A	**HS7067CK**
	+ 10V to + 100V	NSC	8/TO-3	7.0A	**HS7107CK**
	+ 1.25V to + 15V	NSC	2/TO-3	10A	★ **LM396K**
	+ 5V to + 24V	NE	4/TO-204	10A Thermally protected	★ **UA78PGASC**
	± .05V to ± 30V	RAYT	14/DIL	200mA	★ **RC4194DB**

5 pin TO-220

2 pin TO-3

3 pin TO-202

2 pin TO-3

Dual Rail

14 pin DIL

CMOS data sheets

4001B Quad 2 input NOR

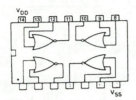

4002B Dual 4 input NOR

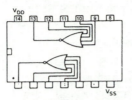

4008B 4 bit full adder

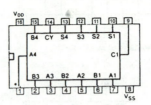

4011B Quad 2 input NAND

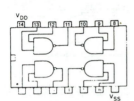

4012B Dual 4 input NAND

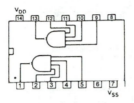

4013B Dual D-type flip-flop

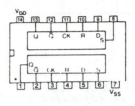

4014B 8 bit shift register

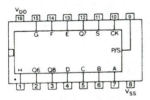

4015B Dual 4 bit shift register

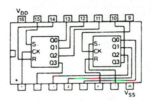

4017B Decade counter divider

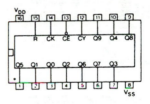

4020B 14 bit binary counter

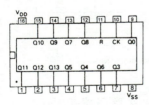

4021B 8 bit shift register

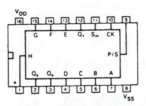

4023B Triple 3 input NAND

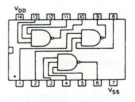

4024B Seven stage ripple counter

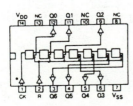

4025B Triple 3 input NOR

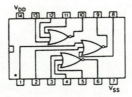

4027B Dual J.K. flip-flop

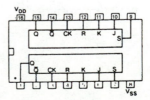

4028B BCD — decimal/binary-octal decoder

V_{DD}

16 15 14 13 12 11 10 9

Q3 Q1 B C D A
Q4 Q8
Q2 Q0 Q7 Q9 Q5 Q6

1 2 3 4 5 6 7 8

V_{SS}

4029B Presettable binary/BCD up/down counter

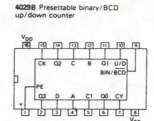

V_{DD}

16 15 14 13 12 11 10 9

CK Q2 C B Q1 U/D
BIN/BCD

PE
Q3 D A C1 Q0 CY

1 2 3 4 5 6 7 8

V_{SS}

4035B 4 bit parallel — in/parallel — out shift register

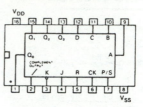

V_{DD}

16 15 14 13 12 11 10 9

Q₁ Q₂ Q₃ D C B

Q_0
COMPLEMENT
OUTPUT
A
K J R CK P/S

1 2 3 4 5 6 7 8

V_{SS}

4040B 12 bit binary counter

V_{DD}

16 15 14 13 12 11 10 9

Q10 Q9 Q7 Q8 R CK Q0
Q11 Q5 Q4 Q6 Q3 Q2 Q1

1 2 3 4 5 6 7 8

V_{SS}

4042B Quad 'D' latch

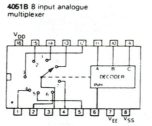

V_{DD}

16 15 14 13 12 11 10 9

Q D CK Q D

1 2 3 4 5 6 7 8

V_{SS}

4043B Quad R/S latch with 3-state outputs "NOR"

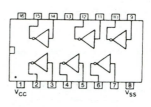

V_{DD}

16 15 14 13 12 11 10 9

R₃ SET 3 SET 2 R₂
Q₃ SET AND RESET Q₂
INPUTS
Q₀ Q₁
R₀ SET 0 EN SET 1 R₁

1 2 3 4 5 6 7 8

V_{SS}

4044B Quad R/S latch with 3-state outputs "NAND"

V_{DD}

16 15 14 13 12 11 10 9

SET 3 R₃ Q₀ R₂ SET 2 Q₂ Q₁
SET AND RESET INPUTS
Q₃ SET 0 R₀ EN R₁ SET 1

1 2 3 4 5 6 7 8

V_{SS}

4047B Monostable/Astable multivibrator

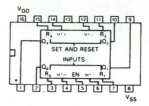

V_{DD}

14 13 12 11 10 9 8

retrigger ÷2 R Q

Ast Ast Mono
Control

1 2 3 4 5 6 7

Timing Enable V_{SS}

4049UB Hex inverter — buffer

V_{DD}

16 15 14 13 12 11 10 9

1 2 3 4 5 6 7 8

V_{CC} V_{SS}

4050B Hex buffer

16 15 14 13 12 11 10 9

1 2 3 4 5 6 7 8

V_{CC} V_{SS}

4051B 8 input analogue multiplexer

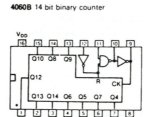

V_{DD}

16 15 14 13 12 11 10 9

A B C
DECODER
INH

1 2 3 4 5 6 7 8

V_{EE} V_{SS}

V_{SS}

4052B Dual 4 input analogue multiplexer

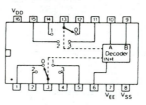

V_{DD}

16 15 14 13 12 11 10 9

A B
Decoder
INH

1 2 3 4 5 6 7 8

V_{EE} V_{SS}

4053B Triple 2 input analogue multiplexer

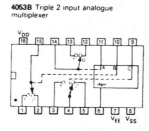

V_{DD}

16 15 14 13 12 11 10 9

A B C

INH

1 2 3 4 5 6 7 8

V_{EE} V_{SS}

4060B 14 bit binary counter

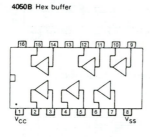

V_{DD}

16 15 14 13 12 11 10 9

Q10 Q8 Q9

Q12
R CK
Q13 Q14 Q6 Q5 Q7 Q4

1 2 3 4 5 6 7 8

V_{SS}

4068B 8-input NAND gate

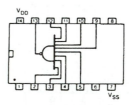

V_{DD}

14 13 12 11 10 9 8

1 2 3 4 5 6 7

V_{SS}

4069UB Hex inverter

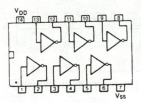

4070B Quad exclusive OR

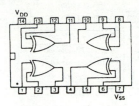

4071B Quad 2 input OR

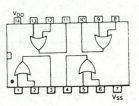

4072B Dual 4-input OR gate

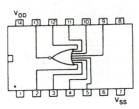

4073B Triple 3 input AND

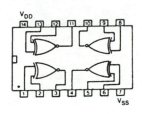

4075B Triple 3 input OR

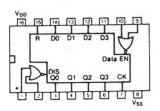

4078B 8 input NOR

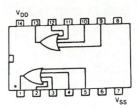

4077B Quad 2 input Exclusive "NOR" gate

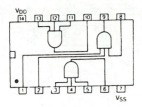

4076B Quad D type register

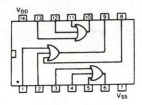

4081B Quad 2 input AND

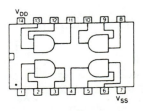

4160B Synchronous programmable 4 bit decade counter with asynchronous clear

4161B Synchronous programmable 4 bit binary counter with asynchronous clear

4162B Synchronous programmable 4 bit decade counter with synchronous clear

4163B Synchronous programmable 4 bit binary counter with synchronous clear

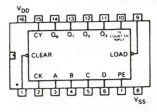

40106B Hex inverting schmitt

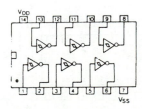

4502B Strobed Hex inverter/buffer

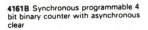
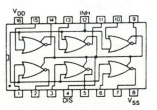

4510B BCD up/down counter

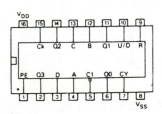

4511B BCD · 7 segment latch/decoder/driver

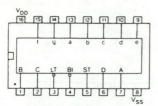

4512B 8 channel data selector

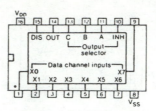

4513B BCD to seven segment latch/decoder/driver

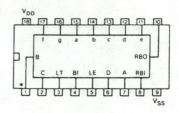

4516B Binary up/down counter

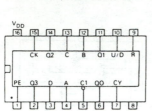

4518B Dual BCD up-counter

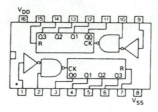

4519B Quad 2 input multiplexer

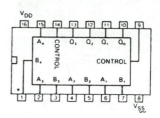

4520B Dual 4 bit binary counter

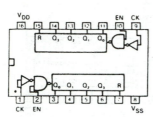

4528B Dual resettable monostable

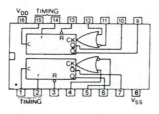

4529B Dual 4-channel analog data selector three state outputs

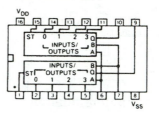

4532B 8 bit priority encoder

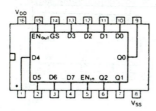

4539B Dual 4 channel data selector/multiplexer

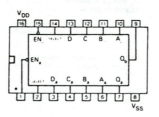

4543B BCD-to-seven segment latch/decoder/driver

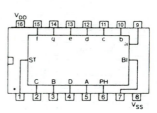

4551B Quad 2-input analog multiplexer/demultiplexer

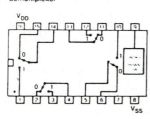

4553B Three-digit BCD counter

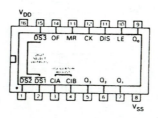

4560B Natural BCD adder

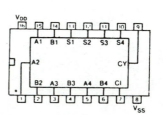

TTL data sheets

00 Quadruple 2-input NAND gate

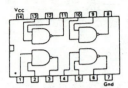

01 Quadruple 2-input NAND gate with open collector output

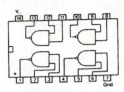

02 Quadruple 2-input NOR gate

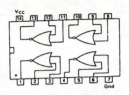

03 Quadruple 2-input NAND gate open collector inputs

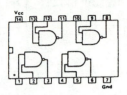

04 Hex inverter. A 74HC004 (un-buffered version) is also available.

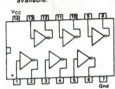

05 Hex inverter-open collector outputs

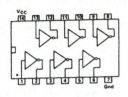

06 Hex inverter with high voltage open collector output

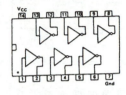

07 Hex driver with open collector output

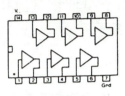

08 Quadruple 2-input AND gate

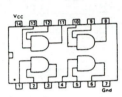

09 Quad 2-input AND gate-open collector outputs

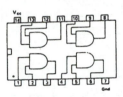

10 Triple 3-input NAND gate

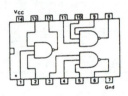

11 Triple 3-input AND gate

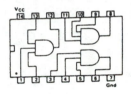

13 Dual 4 input NAND gate Schmitt trigger

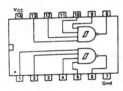

14 Hex Schmitt Trigger

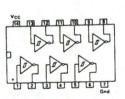

15 Triple 3-input AND gate – open collector outputs

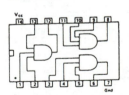

16 Hex Inverter with open collector output

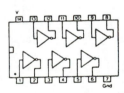

20 Dual 4 input NAND gate

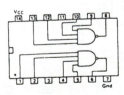

21 Dual 4 input AND gate

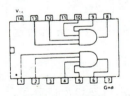

22 Dual 4 input NAND gate open collector outputs

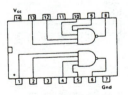

25 Dual 4 input NOR gate with strobe

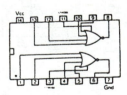

26 Quad 2 input NAND buffer open collector outputs

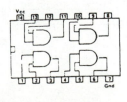

27 Triple 3 input NOR gate

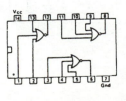

28 Quad 2 input NOR buffer

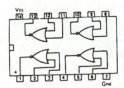

30 8-input NAND gate

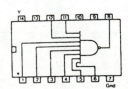

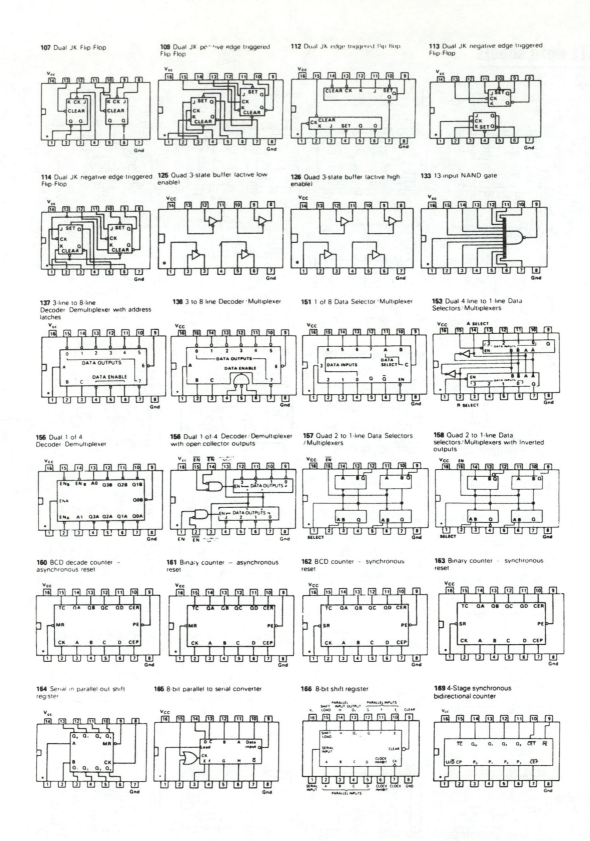

107 Dual JK Flip-Flop

109 Dual JK positive edge triggered Flip Flop

112 Dual JK edge triggered flip flop

113 Dual JK negative edge triggered Flip Flop

114 Dual JK negative edge-triggered Flip-Flop

125 Quad 3-state buffer (active low enable)

126 Quad 3-state buffer (active high enable)

133 13-input NAND gate

137 3-line to 8-line Decoder/Demultiplexer with address latches

138 3 to 8 line Decoder/Multiplexer

151 1 of 8 Data Selector/Multiplexer

153 Dual 4 line to 1-line Data Selectors/Multiplexers

155 Dual 1 of 4 Decoder/Demultiplexer

156 Dual 1-of-4 Decoder/Demultiplexer with open collector outputs

157 Quad 2 to 1-line Data Selectors/Multiplexers

158 Quad 2 to 1-line Data selectors/Multiplexers with inverted outputs

160 BCD decade counter – asynchronous reset

161 Binary counter – asynchronous reset

162 BCD counter – synchronous reset

163 Binary counter – synchronous reset

164 Serial in parallel out shift register

165 8-bit parallel to serial converter

166 8-bit shift register

169 4-Stage synchronous bidirectional counter

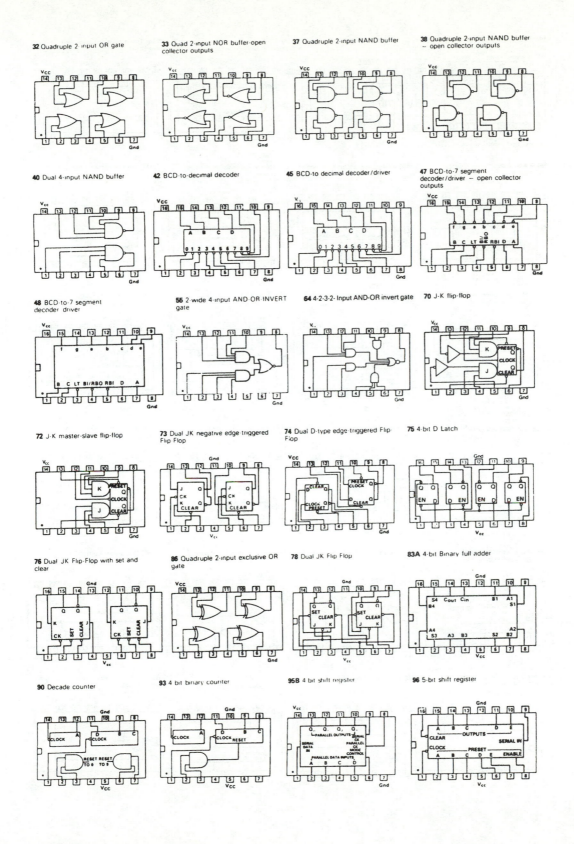

32 Quadruple 2-input OR gate

33 Quad 2-input NOR buffer-open collector outputs

37 Quadruple 2-input NAND buffer

38 Quadruple 2-input NAND buffer — open collector outputs

40 Dual 4-input NAND buffer

42 BCD-to-decimal decoder

45 BCD-to decimal decoder/driver

47 BCD-to-7 segment decoder/driver — open collector outputs

48 BCD-to-7 segment decoder driver

55 2-wide 4-input AND-OR-INVERT gate

64 4-2-3-2- Input AND-OR invert gate

70 J-K flip-flop

72 J-K master-slave flip-flop

73 Dual JK negative edge-triggered Flip Flop

74 Dual D-type edge-triggered Flip Flop

75 4-bit D Latch

76 Dual JK Flip-Flop with set and clear

86 Quadruple 2-input exclusive OR gate

78 Dual JK Flip Flop

83A 4-bit Binary full adder

90 Decade counter

93 4-bit binary counter

95B 4-bit shift register

96 5-bit shift register

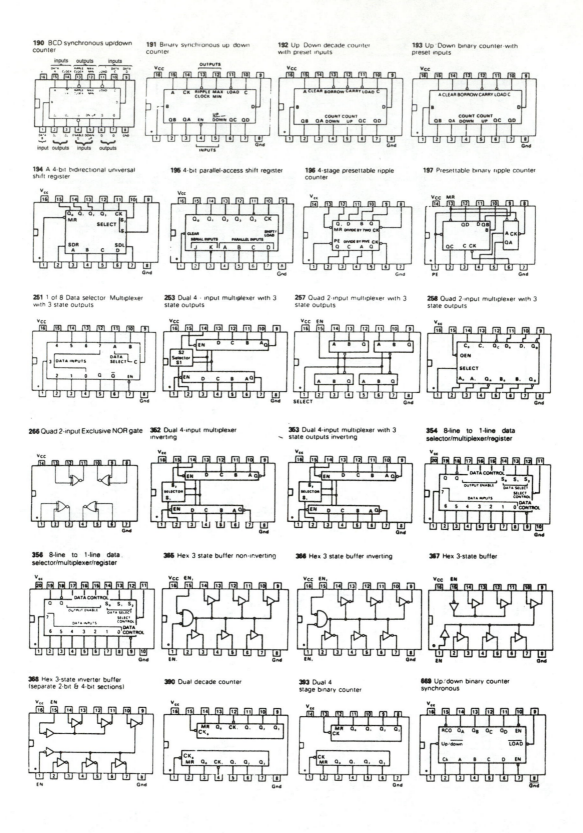

190 BCD synchronous up/down counter

191 Binary synchronous up/down counter

192 Up/Down decade counter with preset inputs

193 Up/Down binary counter with preset inputs

194 A 4-bit bidirectional universal shift register

195 4-bit parallel-access shift register

196 4-stage presettable ripple counter

197 Presettable binary ripple counter

251 1 of 8 Data selector/Multiplexer with 3 state outputs

253 Dual 4-input multiplexer with 3 state outputs

257 Quad 2-input multiplexer with 3 state outputs

258 Quad 2-input multiplexer with 3 state outputs

266 Quad 2-input Exclusive NOR gate

352 Dual 4-input multiplexer inverting

353 Dual 4-input multiplexer with 3 state outputs inverting

354 8-line to 1-line data selector/multiplexer/register

356 8-line to 1-line data selector/multiplexer/register

365 Hex 3 state buffer non-inverting

366 Hex 3 state buffer inverting

367 Hex 3-state buffer

368 Hex 3-state inverter buffer (separate 2-bit & 4-bit sections)

390 Dual decade counter

393 Dual 4 stage binary counter

669 Up/down binary counter synchronous

ANSWERS TO MULTIPLE CHOICE QUESTIONS

—

CHAPTER 1

1: b   2: c   3: c   4: c   5: d   6: b   7: d   8: c

CHAPTER 2

1: c   2: c   3: b   4: a   5: b

CHAPTER 3

1: c   2: d   3: c   4: c   5: b   6: a   7: d   8: d   9: d   10: b

CHAPTER 4

1: b   2: d   3: b   4: d   5: c

CHAPTER 5

1: a   2: c   3: b   4: a   5: d

CHAPTER 6

1: b   2: d   3: a   4: c   5:d

INDEX
